Semiconductor Memory Devices and Circuits

Semiconductor Memory Devices and Circuits

Shimeng Yu

CRC Press
Taylor & Francis Group
Boca Raton London New York

CRC Press is an imprint of the
Taylor & Francis Group, an **Informa** business

First edition published 2022
by CRC Press
6000 Broken Sound Parkway NW, Suite 300, Boca Raton, FL 33487-2742

and by CRC Press
4 Park Square, Milton Park, Abingdon, Oxon, OX14 4RN

CRC Press is an imprint of Taylor & Francis Group, LLC

© 2022 Shimeng Yu

Reasonable efforts have been made to publish reliable data and information, but the author and publisher cannot assume responsibility for the validity of all materials or the consequences of their use. The authors and publishers have attempted to trace the copyright holders of all material reproduced in this publication and apologize to copyright holders if permission to publish in this form has not been obtained. If any copyright material has not been acknowledged please write and let us know so we may rectify in any future reprint.

Except as permitted under U.S. Copyright Law, no part of this book may be reprinted, reproduced, transmitted, or utilized in any form by any electronic, mechanical, or other means, now known or hereafter invented, including photocopying, microfilming, and recording, or in any information storage or retrieval system, without written permission from the publishers.

For permission to photocopy or use material electronically from this work, access www.copyright.com or contact the Copyright Clearance Center, Inc. (CCC), 222 Rosewood Drive, Danvers, MA 01923, 978-750-8400. For works that are not available on CCC please contact mpkbookspermissions@tandf.co.uk

Trademark notice: Product or corporate names may be trademarks or registered trademarks and are used only for identification and explanation without intent to infringe.

Library of Congress Cataloging-in-Publication Data
Names: Yu, Shimeng (Electrical engineer), author.
Title: Semiconductor memory devices and circuits / Shimeng Yu.
Description: First edition. | Boca Raton : CRC Press, 2022. | Includes bibliographical references and index. | Summary: "This book covers semiconductor memory technologies from device bit-cell structures to memory array design with an emphasis on recent industry trends and cutting-edge technologies"-- Provided by publisher.
Identifiers: LCCN 2021053445 (print) | LCCN 2021053446 (ebook) | ISBN 9780367687076 (hardback) | ISBN 9780367687151 (paperback) | ISBN 9781003138747 (ebook)
Subjects: LCSH: Semiconductor storage devices. | Random access memory.
Classification: LCC TK7895.M4 Y85 2022 (print) | LCC TK7895.M4 (ebook) | DDC 004.5--dc23/eng/20211228
LC record available at https://lccn.loc.gov/2021053445
LC ebook record available at https://lccn.loc.gov/2021053446

ISBN: 978-0-367-68707-6 (hbk)
ISBN: 978-0-367-68715-1 (pbk)
ISBN: 978-1-003-13874-7 (ebk)

DOI: 10.1201/9781003138747

Typeset in Times
by SPi Technologies India Pvt Ltd (Straive)

For my wife, Xin Shu, who is fully supportive of my career (including this book write-up) with greatest love.

Contents

Preface xi
Acknowledgments xiii
Author xv

Chapter 1 Semiconductor Memory Technologies Overview 1

    1.1 Introduction to Memory Hierarchy 1
        1.1.1 Data Explosion to Zetta-scale 1
        1.1.2 Memory Hierarchy in the Memory Sub-system 1
    1.2 Semiconductor Memory Industry Landscape 4
    1.3 Introduction to Memory Array Architecture 5
        1.3.1 Generic Memory Array Diagram 5
        1.3.2 Memory Cell Size and Equivalent Bit Area 6
        1.3.3 Memory Array's Area Efficiency 7
        1.3.4 Peripheral Circuits: Decoder, MUX, and Driver 7
        1.3.5 Peripheral Circuits: Sense Amplifier 9
    1.4 Industry Technology Scaling Trend 12
        1.4.1 Moore's Law and Logic Scaling Trend 12
        1.4.2 Definition of Technology Node and Metric for
               Integration Density 13
    1.5 Logic Transistor Technology Evolution 17
    Notes 19
    References 20

Chapter 2 Static Random Access Memory (SRAM) 21

    2.1 6T SRAM Cell Operation 21
        2.1.1 SRAM Array and 6T Cell 21
        2.1.2 Principles of Hold, Read and Write 21
    2.2 SRAM Stability Analysis 27
        2.2.1 Static Noise Margin 27
        2.2.2 N-curve 30
        2.2.3 Dynamic Noise Margin 32
        2.2.4 Read/Write-Assist Schemes 34
    2.3 SRAM's Leakage 37
        2.3.1 Transistor's Sub-threshold Current 37
        2.3.2 SRAM's Leakage Reduction 38
    2.4 Variability and Reliability 39
        2.4.1 Transistor Intrinsic Parameter Fluctuations and the
               Impact on SRAM Stability 39
        2.4.2 Temporal Reliability Issues and the Impact on
               SRAM Stability 41
        2.4.3 Soft Error Caused by Radiation Effects 43

	2.5	SRAM Layout and Scaling Trend 44
	2.5.1	6T Cell Layout 44
	2.5.2	SRAM Scaling Trend 45
	2.6	FinFET-Based SRAM 46
	2.6.1	FinFET Technology 46
	2.6.2	SRAM Scaling in FinFET Era 49

Notes 50
References 51

Chapter 3 Dynamic Random Access Memory (DRAM) 53

- 3.1 DRAM Overview 53
  - 3.1.1 DRAM Sub-system Hierarchy 53
  - 3.1.2 DRAM I/O Interface 53
- 3.2 1T1C DRAM Cell Operation 55
  - 3.2.1 Principle of 1T1C Cell 55
  - 3.2.2 Charge Sharing and Sensing 56
  - 3.2.3 DRAM Leakage and Refresh 59
- 3.3 DRAM Technology 63
  - 3.3.1 Trench Capacitor and Stacked Capacitor 63
  - 3.3.2 DRAM Array Architecture 65
  - 3.3.3 DRAM Layout 66
- 3.4 DRAM Scaling Trend 66
  - 3.4.1 Scaling Challenges 66
  - 3.4.2 Cell Capacitor 68
  - 3.4.3 Interconnect 69
  - 3.4.4 Cell Access Transistor 71
- 3.5 3D Stacked DRAM 74
  - 3.5.1 TSV Technology and Heterogeneous Integration 74
  - 3.5.2 HBM 76
- 3.6 Embedded DRAM 78
  - 3.6.1 1T1C eDRAM 78
  - 3.6.2 Capacitor-less eDRAM 79

Notes 81
References 81

Chapter 4 Flash Memory 83

- 4.1 Flash Overview 83
  - 4.1.1 Flash's History 83
  - 4.1.2 Flash's Applications 84
- 4.2 Flash Device Physics 85
  - 4.2.1 Principle of Floating-Gate Transistor 85
  - 4.2.2 Capacitor Model for Floating-Gate Transistor 86
  - 4.2.3 Program/Erase Mechanism 89
  - 4.2.4 Source-side Injection for Embedded Flash 92

	4.3	Flash Array Architectures	93
		4.3.1 NOR Array	93
		4.3.2 NAND Array	96
		4.3.3 Peripheral Circuits for High Voltage	102
		4.3.4 NAND Flash Translation Layer	103
		4.3.5 Comparison between NOR and NAND	105
	4.4	Multilevel Cell	107
		4.4.1 Multi-level Cell (MLC) Basics	107
		4.4.2 Incremental Step Pulse Programming (ISPP)	109
	4.5	Flash Reliability	111
		4.5.1 Endurance	111
		4.5.2 Retention	113
		4.5.3 Disturb	115
		4.5.4 Trade-offs between Reliability Effects	116
	4.6	Flash Scaling Challenges	117
		4.6.1 Cell-to-cell Interference	117
		4.6.2 Few Electrons Problem	119
	4.7	3D NAND Flash	120
		4.7.1 Principle of Charge-trap Transistor	120
		4.7.2 Cost-Effective 3D Integration Approaches	122
		4.7.3 3D NAND Fabrication Issues	125
		4.7.4 Analysis of the First-generation 3D NAND Chip	126
		4.7.5 New Trends in 3D NAND	128
Notes			131
References			131
Chapter 5	Emerging Non-volatile Memories		133
	5.1	ENVM Overview	133
		5.1.1 Landscape of eNVMs	133
		5.1.2 1T1R Array	138
		5.1.3 Cross-Point Array and Selector	139
	5.2	Phase Change Memory (PCM)	147
		5.2.1 PCM Device Physics	147
		5.2.2 Reliability of PCM	150
		5.2.3 Array Integration of PCM	153
		5.2.4 3D X-point	154
	5.3	Resistive Random Access Memory (RRAM)	155
		5.3.1 RRAM Device Physics	155
		5.3.2 Reliability of RRAM	156
		5.3.3 Array Integration of RRAM	158
	5.4	Magnetic Random Access Memory (MRAM)	159
		5.4.1 MTJ Device Physics	159
		5.4.2 Field Switching MRAM	161
		5.4.3 STT-MRAM	162
		5.4.4 SOT-MRAM	168

- 5.5 Ferroelectric Memories 169
  - 5.5.1 Ferroelectrics Device Physics................................... 169
  - 5.5.2 1T1C FeRAM.. 174
  - 5.5.3 FeFET.. 176
- 5.6 Compute-in-Memory.. 180
  - 5.6.1 Principle of CIM... 180
  - 5.6.2 Synaptic Device Properties 182
  - 5.6.3 CIM Prototype Chips 185
- Notes.. 188
- References 188

Index 195

Preface

The emerging applications enabled by artificial intelligence (AI) are becoming indispensable in our daily life. AI software models demand efficient hardware technologies to store and process an enormous amount of data. This book aims to cover the fundamentals of semiconductor memory devices and peripheral circuits and the recent industry trends in scaling the memory cells toward the sub-10 nm scale and vertical integration via three-dimensional (3D) stacking.

The functionality and performance of today's computer architectures are increasingly dependent on the characteristics of the components in the memory hierarchy. The memory hierarchy includes on-chip cache and off-chip standalone memories such as main memory, storage-class-memory, and solid-state drive. This book will introduce the semiconductor memory technologies that serve various levels of the memory hierarchy, from the device cell structures to the array-level design, with an emphasis on the recent industry trend and cutting-edge technologies. Since this is a rapidly evolving field, many discussions in this book are based on the state-of-the-art as of 2020, with reasonable projections toward the next decade.

Chapter 1 will present an overview of the semiconductor memory technologies, including the concept of the memory hierarchy, the generic memory array diagram and common peripheral circuit modules, metrics for evaluating the bit density, and array area efficiency.

Chapters 2–4 will present the mainstream semiconductor memory technologies, static random access memory (SRAM), dynamic random access memory (DRAM), and Flash memory, respectively. Topics such as basic operation principles, device physics, manufacturing processes, bit-cell design considerations, array architectures, and technology scaling challenges will be discussed. Recent industry trends such as FinFET-based SRAM, high-bandwidth memory (HBM), and 3D vertical NAND Flash will be introduced.

Chapter 5 will survey the emerging non-volatile memory (eNVM) candidates that may have the potential to change the memory hierarchy, or serve as embedded memories on-chip and enable new applications beyond data storage. Candidates of interest include the phase change memory (PCM) and the related selector for 3D X-point technology, resistive random access memory (RRAM), magnetic random access memory (MRAM) including spin-transfer-torque (STT), and spin-orbit-torque (SOT) switching mechanism, as well as ferroelectric memories such as ferroelectric random access memory (FeRAM) and ferroelectric field-effect transistor (FeFET). The multilevel capability, variability, and reliability issues of eNVM devices will be addressed. Chapter 5 also introduces the concept of the compute-in-memory that merges the mixed-signal computation into the memory arrays for accelerating the vector-matrix multiplication in the deep neural networks, which is an attractive paradigm for machine learning hardware accelerator.

The materials covered in this book are based on the graduate-level course that the author has been offering eight times in the past decade at the Arizona State University and the Georgia Institute of Technology. More than 500 students have taken this course and many of them are now working in the semiconductor industry. The author has also released his recent recorded lectures on YouTube publicly, which has attracted more than 20,000 views. This book will serve as a textbook for graduate-level curriculum in electrical and computer engineering programs, and/or a reference for entry-level industry engineers and researchers.

Acknowledgments

The author would like to thank his students, Anni Lu, Hongwu Jiang, Dong Suk Kang, Yuan-chun Luo, Yandong Luo, and his postdoc researcher Dr. Wonbo Shim for their help in graphics editing and text proofreading.

Author

Shimeng Yu Shimeng Yu is currently an associate professor of electrical and computer engineering at the Georgia Institute of Technology. He received a B.S. degree in microelectronics from Peking University in 2009, and M.S. degree and Ph.D. degree in electrical engineering from Stanford University in 2011 and 2013, respectively. From 2013 to 2018, he was an assistant professor at the Arizona State University.

Prof. Yu's research interests include semiconductor devices and integrated circuits for energy-efficient computing systems. His research expertise is on emerging nonvolatile memories for applications such as deep learning accelerator, in-memory computing, 3D integration, and hardware security. He has published more than 350 peer-reviewed conference and journal papers and with more than 18,000 Google Scholar citations with an H-index 64.

Among Prof. Yu's honors include being a recipient of NSF Faculty Early CAREER Award in 2016, IEEE Electron Devices Society (EDS) Early Career Award in 2017, ACM Special Interests Group on Design Automation (SIGDA) Outstanding New Faculty Award in 2018, Semiconductor Research Corporation (SRC) Young Faculty Award in 2019, ACM/IEEE Design Automation Conference (DAC) Under-40 Innovators Award in 2020, IEEE Circuits and Systems Society (CASS) Distinguished Lecturer for 2021–2022, IEEE Electron Devices Society (EDS) Distinguished Lecturer for 2022–2023.

Prof. Yu has been serving many premier conferences as part of the technical program committees, including IEEE International Electron Devices Meeting (IEDM), IEEE Symposium on VLSI Technology and Circuits, IEEE International Reliability Physics Symposium (IRPS), ACM/IEEE Design Automation Conference (DAC), ACM/IEEE Design, Automation & Test in Europe (DATE), ACM/IEEE International Conference on Computer-Aided Design (ICCAD), etc. He is an editor of the *IEEE Electron Device Letters*. He is also a senior member of the IEEE.

1 Semiconductor Memory Technologies Overview

1.1 INTRODUCTION TO MEMORY HIERARCHY

1.1.1 Data Explosion to Zetta-scale

Digital data volume is exploding. A recent analysis [1] predicted that the number of devices connected via the Internet could reach almost 75 billion globally by 2025. Furthermore, the data generated from these devices will reach 175 Zettabytes[1] by 2025, with most of it coming from videos and security camera surveillance. Therefore, analyzing the data and finding ways for the short-term and long-term storage of the data are essential. Archiving these vast volumes of data is also important. New data are generally collected at the edge, partially processed at the edge, partially transmitted to the cloud, and mostly stored at the cloud. The ever-increasing demand for data analytics and data storage drives the continuing development of memory/storage technologies for higher density and wider bandwidth. A rough differentiation factor between memory and storage is the lifetime of the data.[2] The memory holds short-term data with faster and more frequent read/write access, while the storage holds long-term data with slower and less frequent read/write access.

1.1.2 Memory Hierarchy in the Memory Sub-system

A modern computer system generally follows the von-Neumann architecture where the data are processed in the processor core (e.g., arithmetic logic unit, ALU) and the data are stored in the memory components. Ideally, a memory device should have a sufficiently large capacity and fast access speed. Unfortunately, there exists no such "universal" memory that can satisfy both needs. Therefore, different memory components are required to build the memory sub-system, establishing a memory hierarchy. The memory hierarchy is typically shown as a pyramid in Figure 1.1. Going upward the hierarchy, the memory component is becoming faster; going downward the hierarchy, the memory component has a larger capacity.

The top of the pyramid is the processor core, and the core typically has the logic units and the cache integrated on the same chip (as indicated by the box). Cache stores the most frequently used data, and it is typically composed of multi-levels (L1, L2, L3, and/or the last level) as well. Static random access memory (SRAM) is the primary on-chip memory technology that implements L1 to L3 cache. A trade-off

DOI: 10.1201/9781003138747-1

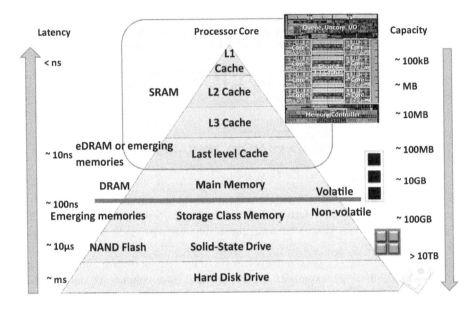

FIGURE 1.1 The memory hierarchy showing mainstream SRAM-based cache, DRAM-based main memory, and NAND Flash-based solid-state drive, and opening gaps for eDRAM or emerging memories to serve the last level cache and/or storage class memory.

between the access latency and the storage capacity is also present within the cache levels. For example, the L1 cache could be accessed in sub-ns and has a capacity of ~100 kB,[3] the L2 cache could be accessed in 1–3 ns and has a capacity of ~1 MB, and the L3 cache could be accessed in 5–10 ns and has a capacity of tens of MB. In some high-performance computing systems, for instance, IBM's power series microprocessor [2], the last level cache is implemented with an embedded dynamic random-access memory (eDRAM).

Off the processor's chip, the main memory in the hierarchy is generally implemented by the standalone DRAM,[4] which could be accessed in tens of ns and has a capacity of tens of GB. Most of the data that is readily used by a software program is stored in the main memory. Both SRAM-based cache and DRAM-based main memory are classified as volatile memories, which means that the data will not be preserved when the power supply is removed. Sometimes, they are also referred to as working memories. If the data need to be preserved for a long time even when the power supply is removed, non-volatile memories (NVMs) are required as the storage memories. The line indicating the boundary between the volatile memories and the NVMs is shown in the pyramid. The widely used NVMs are the NAND Flash-based solid-state drive (SSD) and the magnetic hard-disk drive (HDD). SSD could be accessed in tens of μs and has a capacity of hundreds of GB–TB, and HDD could be accessed in ~ms and has a capacity of tens of TB. The perceived and ever-increasing gap between the main memory's bandwidth and the SSD's bandwidth motivates a recent trend of creating a new level in the memory hierarchy, namely the storage class memory which could be accessed in hundreds of ns and has a capacity of tens

Semiconductor Memory Technologies Overview

to hundreds of GB. The storage class memory is placed at the boundary between the working memories and the storage memories and very often it belongs to NVMs, thus sometimes it is also referred to as the persistent memory. Emerging memories are actively being researched to fill in this vacant position as storage-class memory, and a notable example is the three-dimensional (3D) X-point memory introduced by Intel and Micron [3]. Figure 1.2 shows the trade-offs between access time, integration density (Mb/mm^2), and cycling endurance of different memory technologies in the memory hierarchy. The recent industrial trends in 3D vertical NAND Flash and 3D stacked DRAM are also shown. The necessity of creating a storage class memory to bridge the gap between NAND Flash and DRAM is indicated, which opens opportunities for emerging memories such as resistive random access memory (RRAM) and phase-change memory (PCM). To bridge the gap between DRAM and SRAM, emerging memories such as magnetic random access memory (MRAM) are suited for the last-level cache. The newest technology that may find its position in this chart is the ferroelectric field-effect transistor (FeFET). A 3D stacked high-density FeFET may serve as storage-class memory, and a FeFET with optimized write endurance may serve as the last level cache. Another important aspect of the trade-off in this

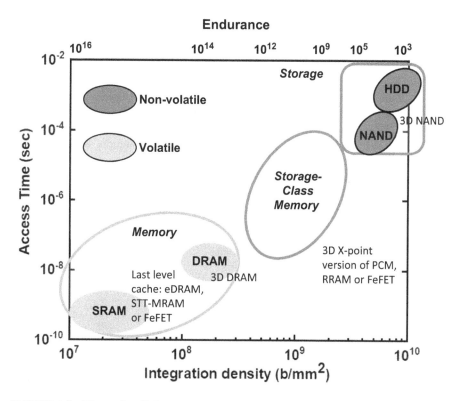

FIGURE 1.2 The trade-offs in access time, integration density, and cycling endurance for different memory technologies, indicating the opportunities for emerging memories in the storage-class memory and the last level cache.

chart is the cycling endurance, which specifies how many times the memory device could be written before it fails. As working memories, SRAM or DRAM generally have >10^{16} endurance in a 10-year lifetime. NAND Flash is less often written with 10^3–10^5 endurance, and the storage-class memory is expected to have 10^9–10^{12} endurance.

There are other storage media beyond the technologies listed in this pyramid. For archiving purposes, magnetic tapes are still being used for the large volume "cold" data storage due to their ultra-low cost. Optical discs such as CDs, DVDs, or Blu-ray are also used for information distribution purposes. The magnetic tapes, optical discs, and HDD will not be covered in this book as they are not fabricated with the silicon-manufacturing processes. To summarize, the focus of this book will be placed on the semiconductor memory technologies such as SRAM, DRAM, NAND Flash, and emerging memories.

1.2 SEMICONDUCTOR MEMORY INDUSTRY LANDSCAPE

As introduced in the memory hierarchy, the semiconductor memories could be categorized into two types: the embedded memories that are on the same processor's chip and the standalone memories that are off the processor's chip. As of 2020, the total semiconductor revenue is $466 billion, out of which $126 billion were contributed by the standalone memories [4]. 98% of revenue for standalone memories came from DRAM (53%) and NAND Flash (45%). Figure 1.3 shows the market share of DRAM and NAND Flash as of 2020, with respect to the total semiconductor revenue. The standalone memories industry has consolidated over the decades and now it is dominated by a few key vendors. The major DRAM vendors are Samsung, Micron, and SK Hynix, and the major NAND Flash vendors are Samsung, Micron, SK Hynix, Kioxia (a spin-off from Toshiba in 2018), Western Digital (who acquired SanDisk in 2016), and Intel (who is scheduled to sell its memory business to SK Hynix by 2025).

The embedded memories are components of a microprocessor (e.g., central processing unit, CPU or graphics processing unit, GPU); therefore, it is more challenging to estimate its market share.[5] It was estimated that ~$35 billion revenue was

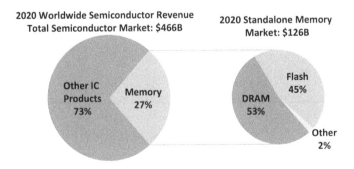

FIGURE 1.3 Semiconductor revenue (2020) and its distribution into standalone memories (DRAM and NAND Flash).

generated by the embedded memories components as of 2020, mostly contributed from the SRAM for microprocessor or microcontroller,[6] and partly contributed from the embedded NVMs for the microcontroller. Foundries are the primary vendors of the embedded memories, and the major players are Taiwan Semiconductor Manufacturing Corporation (TSMC), Samsung, Globalfoundries, etc. SRAM scales well with the logic transistor process to the leading-edge node, but the conventional embedded Flash (eFlash) is difficult to scale beyond the 28 nm node. Therefore, foundries are actively researching emerging NVMs (eNMVs) that could be extended to more advanced nodes.

1.3 INTRODUCTION TO MEMORY ARRAY ARCHITECTURE

1.3.1 Generic Memory Array Diagram

Regardless of the types of semiconductor memory technologies, they are commonly organized into memory sub-arrays. Figure 1.4 shows an example of the generic memory sub-array diagram with a two-dimensional (2D) matrix.[7] In the case of a 2D matrix, the rows are typically called word lines (WLs), and the columns are typically called bit lines (BLs). The cross-point intersection of the BL and the WL is a memory cell. The memory cells defined by WLs and BLs form the memory cell array. For any memory technology to be functional, appropriate peripheral circuits are indispensable in the sub-array. The common peripheral circuits modules include the row decoder which translates the row address M-bit to select one of the 2^M rows, the column decoder which translates the column address N-bit to select one of the 2^N column groups via the column multiplexer (MUX), the read sense amplifier, and the write driver. Here, one column group contains k columns, and there is no further decoding within one column group thus k bits are transferred in parallel to the sense amplifier or from the write driver. Hence, the input/output (I/O) width from the sub-array is k-bit. In total, there are 2^M rows and $k \times 2^N$ columns; thus, $k \times 2^M \times 2^N$ cells

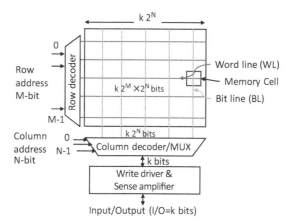

FIGURE 1.4 The generic memory sub-array diagram organized in a 2D matrix, showing the memory cell array and its peripheral circuit modules.

are in the memory cell array. Typically, when the memory state is being read out, it generates a small amplitude analog signal, which needs to be sensed and amplified by the sense amplifier to the digital 0 and 1 as output. The write driver is also needed to charge up the WL/BL when the data are written into the memory array from the external data bus.

Trade-offs exist in sub-array size among different levels in the memory hierarchy. For fast access, the sub-array size is typically small. For instance, SRAM sub-array size ranges from 32 × 32, 64 × 64, 128 × 128 to 256 × 256 for L1 to L3 caches, and DRAM sub-array size is moderately sized up (e.g., 512 × 512, 1024 × 1024). On the other hand, for large-capacity storage, NAND Flash's sub-array size could be as huge as 16k × 64 to 128k × 64.

1.3.2 Memory Cell Size and Equivalent Bit Area

Depending on the specific type of the semiconductor memory technologies, the actual implementation of the memory cell is different in the generic memory sub-array diagram. The SRAM cell typically consists of six logic transistors, the DRAM cell typically consists of one access transistor and one capacitor, and the Flash memory cell typically consists of a single floating-gate transistor or charge-trap transistor. The integration density varies significantly among the memory technologies, and a higher integration density means a lower cost per bit. Lower cost per bit is essential to achieve larger storage capacity given a certain form factor of the chip. As a widely used metric to evaluate the memory cell size, F^2 is used. Here, F is the feature size of a particular technology node, thus F^2 could be regarded as the minimum unit area in a certain technology node.

Figure 1.5 shows the normalized equivalent bit area in terms of F^2 for a variety of semiconductor memory technologies. SRAM is typically in the range of 150–300 F^2 (in most of the technology nodes down to 22 nm). Depending on at which level the cache is, the SRAM transistor width may be sized accordingly. L3 cache typically

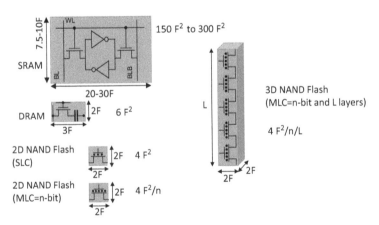

FIGURE 1.5 The normalized equivalent bit area in terms of F^2 for a variety of semiconductor memory technologies.

Semiconductor Memory Technologies Overview

employs the high-density SRAM cells with minimum transistor width, leading to the lower bound of ~150 F^2. On the other hand, the L1 cache requires the fastest access. Hence, the transistor width is sized up for providing more drive current, resulting in the higher bound of ~300 F^2. DRAM has a very compact layout design rule that pushes the cell size to be 6 F^2 (in most of the technology nodes). Apparently, DRAM has a much higher integration density (thus larger capacity) than SRAM. It should be noted that both SRAM and DRAM are binary memories that each memory cell could only store 1 bit (digital 0 or 1). Therefore, their memory cell size is the same as the equivalent bit area.

Flash memory is more complicated as it offers multilevel cell (MLC) capability, which means one memory cell could store n bits ($n \geq 2$), and n ranges from 2 to 3, 4, and possibly even more.[8] In a 2D NAND Flash, the typical memory cell size is 4 F^2. If it is a single-level cell (SLC), the equivalent bit area is also 4 F^2. If it is an MLC, the equivalent bit area is reduced to 4 F^2/n. The advanced 3D NAND Flash today has stacked many layers vertically on the same 2D footprint to further improve the integration density. If the 3D stacked layer number is L (L could be 144 or 176 as of 2020), then the equivalent bit area is aggressively scaled to $4F^2$/n/L for n-bit per cell. Therefore, the integration density of 3D NAND Flash with MLC is superior to any other semiconductor memory technologies, resulting in the lowest cost per bit.

1.3.3 Memory Array's Area Efficiency

Another important metric to evaluate the chip-level integration density is the array's area efficiency, which is defined as the area percentage occupied by the memory cell array over the entire memory chip area. The entire memory chip area includes the memory cell array and the peripheral circuits. The higher the area efficiency, the lower the cost per bit. This is because the peripheral circuits aid in decoding, sensing, programming the memory cells but they do not directly contribute to the information storage. The valuable silicon footprint is preferred to be spent on the memory cell array. Therefore, it is desirable to minimize the area occupied by the peripheral circuits for a higher chip-level integration density. Generally, NAND Flash has the highest area efficiency 70–80%. DRAM has a moderate area efficiency of 60–70%. SRAM has the lowest area efficiency ~50% or below. Figure 1.6 shows examples of die photos of a NAND Flash and an SRAM cache (embedded with a microprocessor). It should be pointed out that when calculating a certain level of SRAM cache's area efficiency, the other logic core area is not counted as the total memory chip area. The memory chip area should be specified to a certain level of cache only (e.g., the L3 block area highlighted in the die photo for estimating the L3's area efficiency).

1.3.4 Peripheral Circuits: Decoder, MUX, and Driver

For decoding the bit address, the digital decoder is used. If the bit address is long (e.g., M > 5 bit), it is not area-efficient to apply the direct decoding from M-bit to 2^M rows. In this case, a two-stage decoder is used. For example, the 8-bit address is first decoded to $2^4 = 16$ intermediate nodes, and each node is decoded to another $2^4 = 16$ rows. An example of such an 8-to-256 two-stage decoder is shown in Figure 1.7(a).

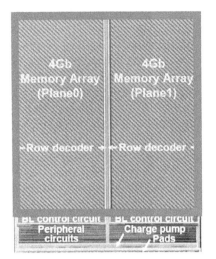
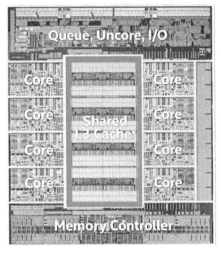

FIGURE 1.6 Examples of die photos of a NAND Flash and an SRAM L3 cache (embedded with a microprocessor).

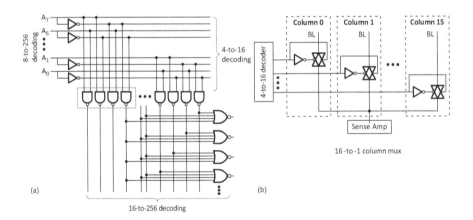

FIGURE 1.7 (a) An example of 8-to-256 two-stage decoder; (b) an example of 16-to-1 MUX that is controlled by a 4-to-16 decoder.

The decoder design applies to both row decoding and column decoding. The column selection needs a MUX that is controlled by the column decoder. The MUX typically employs transmission gates. Figure 1.7(b) shows an example of a 16-to-1 MUX that is controlled by a 4-to-16 decoder (which could be the second stage of the two-stage decoder). The transmission gate's transistor width may need to be sized accordingly to deliver sufficient write current to the memory cell.

After the row decoder, there is typically a WL driver to drive the long wires with parasitic capacitance. Figure 1.8 illustrates a few design options for the WL driver with various optimization goals. In this example, the lumped WL capacitance is assumed to be 4096 unit capacitance,[9] thus the driver could be divided into stages to

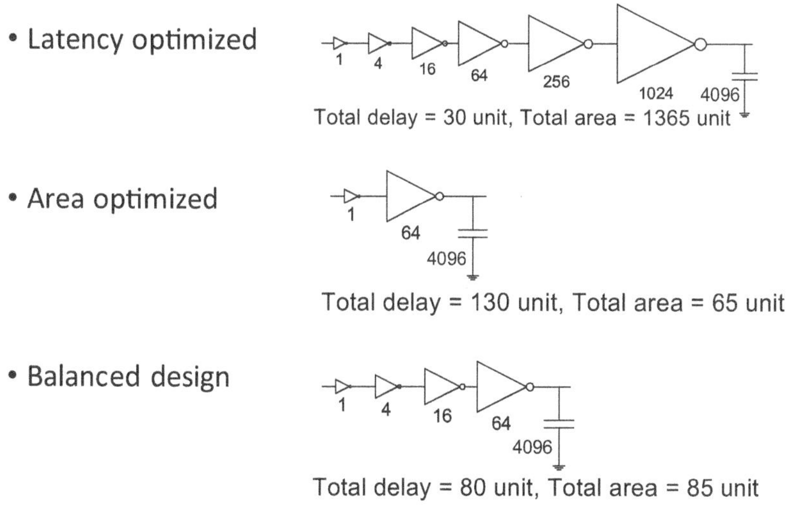

Semiconductor Memory Technologies Overview

- Latency optimized

  Total delay = 30 unit, Total area = 1365 unit

- Area optimized

  Total delay = 130 unit, Total area = 65 unit

- Balanced design

  Total delay = 80 unit, Total area = 85 unit

FIGURE 1.8 Design options for the WL driver with various optimization goals: (a) latency optimized; (b) area optimized; (c) balance between latency and area.

trade-off between the latency and the area. If the optimization goal is minimizing the latency, the number of stages and the inverter sizing in the inverter chain could be calculated by the logical effort, resulting in a 6-stage design with increasing inverter size as shown in Figure 1.8(a). In this case, the driver delay is minimized to 30 unit delay but the driver area is as large as 1365 unit area.[9] If the optimization goal is minimizing the area, the number of stages could be just 2, as shown in Figure 1.8(b). In this case, the driver delay is increasing to 130 unit delay but the driver area is as small as 65 unit area. A more balanced design could determine the size of the last stage inverter by its driving current requirement and then calculate the size of the other inverters by logical effort, as shown in Figure 1.8(c). In this case, the driver delay is reduced to 130 unit delay but the driver area is slightly enlarged to 85 unit area. Depending on the specific memory type, the optimization goal varies. For SRAM caches, minimizing latency is of higher importance, while for NAND Flash, minimizing the area overhead of peripheral circuits is preferred.

1.3.5 Peripheral Circuits: Sense Amplifier

A sense amplifier (SA) is one of the most important peripheral circuit modules that converts the analog small signal to digital output (0 or 1) during the read operation. Figure 1.9 shows the bit line resistance–capacitance (RC) model that models the wire resistance and parasitic capacitance in a distributed network. The memory cell is modeled as a current source (I_m) with a parallel output resistor (R_m), and the input stage of SA is modeled as a load resistor (R_L). Using the transmission line theory, the propagation delay (Δt) could be calculated as

$$\Delta t = \frac{R_T C_T}{2}\left(\frac{R_m + R_T/3 + R_L}{R_m + R_T + R_L}\right) + R_m C_T \left(\frac{R_L}{R_m + R_T + R_L}\right) \quad (1.1)$$

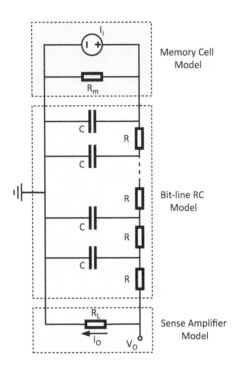

FIGURE 1.9 The bit line RC model with a distributed network.

where the R_T is the total wire resistance, and C_T is the total wire parasitic capacitance. The SA design is generally divided into two modes: voltage sensing and current sensing.

If it is the voltage sensing, the input resistance of SA could be regarded as infinity as an open-circuit load, therefore, Equation (1.1) could be approximated as follows:

$$\Delta t = \frac{R_T C_T}{2}\left(1 + \frac{2R_m}{R_T}\right) \qquad (1.2)$$

If it is the current sensing, the input resistance of SA could be regarded as zero as a short-circuit load, therefore, Equation (1.1) could be approximated as follows,

$$\Delta t = \frac{R_T C_T}{2}\left(\frac{R_m + R_T/3}{R_m + R_T}\right) \qquad (1.3)$$

Depending on the exact values of the memory cell resistance, wire resistance, and parasitic capacitance, the latency between voltage mode SA and current mode SA could be compared using the above equations as a back-of-envelop calculation.

The transistor-level implementations of the voltage mode SA and the current mode SA can have various circuit topologies. Figure 1.10 shows some examples of voltage mode SA and current mode SA implementations that can serve the resistive memories[10] that use a high resistance state (HRS) and a low resistance state (LRS) to

Semiconductor Memory Technologies Overview

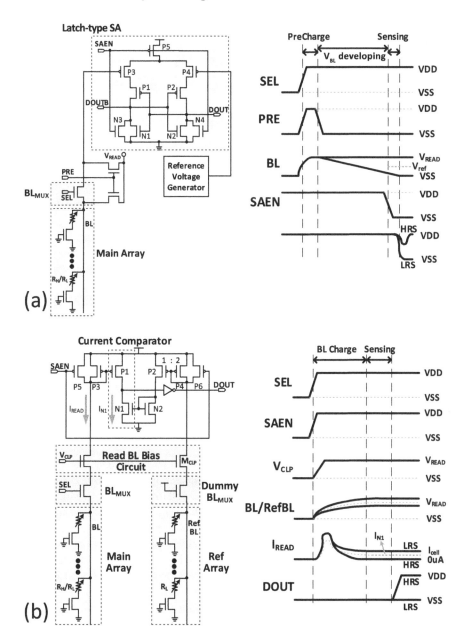

FIGURE 1.10 Examples of (a) voltage mode sense amplifier and (b) current mode sense amplifier.

store the information "0" and "1" [5] For the voltage mode SA in Figure 1.10(a), BL is pre-charged to the read voltage V_{READ} and both the output nodes (DOUT and DOUTB) are grounded by a high SAEN signal (as the latch M1–M4 is disconnected from the power supply). Then column MUX is turned on by Y_{SEL} and the BL voltage decays through the pull-down path through the memory cell. If the memory cell is in

HRS, the pull-down current is small, and the BL decay is slow; if the memory cell is in LRS, the pull-down current is large, and the BL decay is fast. After a BL voltage development time, a BL voltage window is opened up between HRS and LRS sensing. Now the SA is enabled by a low SAEN signal as this SA has a PMOS differential input pair. The BL voltage is compared against the reference voltage as the gate input to the p-type differential pair (M5 and M6). The latch (M1–M4) will flip. If the BL voltage is higher/lower than the reference voltage for sensing HRS/LRS, DOUT will flip to ground/V_{DD}, respectively. For the current mode SA in Figure 1.10(b), SAEN is low before the column MUX Y_{SEL} is enabled. When Y_{SEL} is turned on at the same time, the clamp transistor (M_{CLP}) gate voltage rises to the read voltage V_{READ} and maintains a constant bias during the current-mode sensing. The BL current will see an initial peak as being charged by M1, but it will decay over time when BL is discharged by the pull down paths through the memory cells. If the memory cell is in HRS, the pull-down current is small. If the memory cell is in LRS, the pull-down current is large. Then the BL current will be transferred to M5 by the current mirror of M1/M3. A smaller/larger current in M5 when sensing HRS/LRS is compared against the reference current flowing in M6, which will cause a higher/lower voltage than the middle voltage of the output inverter driver, and then flip DOUT to be ground/V_{DD}, respectively. The general comparison results between these two modes indicate that for sensing smaller current (from a higher memory cell resistance) or longer BL (with larger parasitic capacitance), current-mode SA is preferred as it could have a smaller delay.

1.4 INDUSTRY TECHNOLOGY SCALING TREND

1.4.1 Moore's Law and Logic Scaling Trend

It was predicted by Dr. Gordon Moore, Intel's co-founder, in the late 1960s, that the transistor count on a processor chip (or a die) would exponentially grow (i.e., double approximately every 2 years. This prediction is referred to as the famous Moore's law [6], and it has become the driving force for the technology scaling in the semiconductor industry over the past 5 decades. From generation to generation, the effective area per transistor is supposed to shrink by ~0.5×, which is translated to a dimensional scaling by ~0.7× (the square root of 0.5×). It should be noted that Moore's law is not a physical law that defines the transistor's dimension must be downscaled by 0.7×, but rather is an economic law that aims to lower the cost per transistor. At the same time, better performance or more chip functionalities are obtained as a byproduct of the scaling. Figure 1.11 plots the transistor count on the die for a variety of microprocessors (including Intel's personal computer/server CPUs, Apple's mobile CPUs, and Nvidia's GPUs), as well as DRAM and NAND Flash. A few observations could be made from this chart. First, the microprocessors today generally contain more than a billion transistors on the die (and a big portion of these transistors are used to build SRAM). Second, the transistor count of standalone memories (DRAM and Flash) on a die outpaces that of the microprocessors in the recent decade. The NAND Flash (especially the 3D version of it) is the technology that has the highest integration density, approaching a trillion of transistors on the die.

Semiconductor Memory Technologies Overview

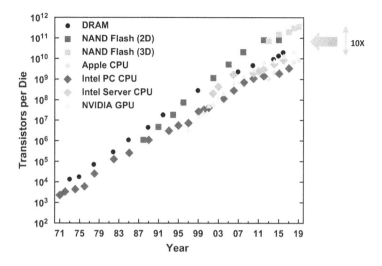

FIGURE 1.11 Moore's law showing the transistor count on the die for a variety of processors as well as DRAM and NAND Flash over the past 5 decades.

It has been suggested that Moore's law is slowing down since the mid-2010s, and the traditional 2D scaling will soon approach its fundamental limit in the mid-2020s as the physical space is run out [7]. However, new opportunities toward the third dimension (3D) are on the horizon. For instance, heterogeneous and/or monolithic 3D integration may continue to improve the integration density, lower the cost per transistor (or per bit), and further boost the system-level performance and/or add on new functionalities. Memory technologies including DRAM and NAND Flash have been among the first to adopt these new 3D integration schemes, which will be discussed in more detail in Sections 3.5 and 4.7, respectively.

1.4.2 Definition of Technology Node and Metric for Integration Density

The 2D technology scaling used to be reflected as the downscaling of the technology node. The technology node F is often regarded as the minimum lithographic feature size, which is true for DRAM and NAND Flash. Figure 1.12 shows the definition of F being the half-pitch between the M1 (the first metal layer) wires for DRAM or the half-pitch between the poly-silicon gates for 2D NAND Flash. In the 3D vertical NAND structure, the F is usually measured as the half-pitch of the center-to-center distance from one pillar to the neighboring pillar.

However, it is tricky to correlate the F to any physical dimensions of the logic process (including the SRAM cell that uses logic transistors). Figure 1.13 shows the trend of silicon logic process scaling featuring the gate length and the technology node name. Historically, the technology node was the same as the gate length of the logic transistor before the mid-1990s. But this is no longer true below 350 nm node, where the gate length was scaled aggressively to be much smaller than the technology node (for the several generations like 250 nm, 180 nm, 130 nm, 90 nm, and 65 nm).

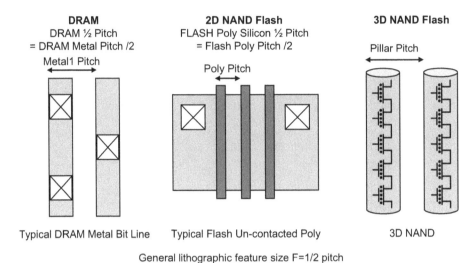

FIGURE 1.12 The definition of lithographic feature size F for DRAM, 2D NAND Flash, and 3D NAND Flash.

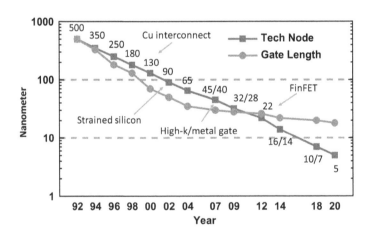

FIGURE 1.13 The trend of silicon logic process scaling featuring the gate length and the technology node name in the past 3 decades. Key technological innovations are indicated.

Then in the late 2000s, the trend became reversed where the scaling of the gate length slowed down (due to limitations imposed by the short channel effects). The industry also diverged in naming the technology node. For instance, Intel's 45 nm node is equivalent to TSMC's 40 nm node, and Intel's 32 nm node is equivalent to TSMC's 28 nm node. After that, the technology node name keeps shrinking following the 0.7× rule as 22 nm in 2012, 14 nm in 2014, 10 nm in 2016, 7 nm in 2018, 5 nm in 2020, and 3 nm in 2022. Now the technology node no longer represents any physical dimension on the transistor structure, because the gate length barely scales below the

Semiconductor Memory Technologies Overview

22 nm node. Even for a 5 nm node logic transistor, the gate length is physically ~20 nm. Now, it is to better view the technology node name as a symbolic representation of a certain technology generation [8]. In this book, the SRAM cell that uses the logic process still assumes that F is the same as the technology node simply for a normalization purpose, thus it should be treated with caution especially in the leading-edge nodes.

Now the question becomes what is really scaling since the gate length no longer scales. To solve the ambiguity and better capture the nature of scaling, an alternative metric to evaluate the logic transistor density is proposed by using the contacted gate pitch (CGP)[11] and the M1 pitch. Figure 1.14's inset shows a simplified top-view layout of a FinFET-based logic standard cell. The CGP is the horizontal distance between the centers of the two adjacent gate contacts that are separated by a source/drain contact via. The M1 pitch determines the vertical dimension. Therefore, the effective area for one transistor is the product of the CGP and the M1 pitch. Since the gate length no longer scales today, what really scales is the CGP (with possibly reduced spacer distance between the gate and the source/drain contact vias). Figure 1.14 shows the scaling trend of the CGP and the M1 pitch below 45 nm in the past decade and the projection into 3 nm, 2 nm, and 1 nm nodes, indicating a straight downscaling trend with the slope to be extracted to be ~0.75× per generation till 7 nm node. As mentioned earlier, different companies have their own strategies in naming the node. Figure 1.15 shows the comparison between Intel and TSMC in the CGP and M1 dimensions in their recent nodes. It is seen that Intel's 10 nm process is approaching TSMC's 7 nm process in terms of integration density.

In the non-planar FinFET era in the late 2010s, it had become feasible to effectively increase the drive current by increasing the fin height. With sufficiently high drive current, it is possible to perform fin depopulation, i.e., eliminating fins which enable metal tracks reduction in the complementary-metal-oxide-semiconductor

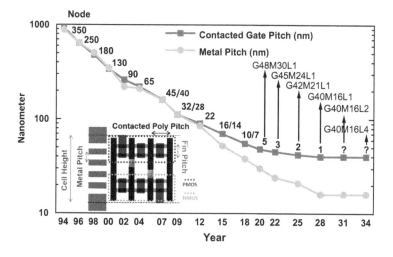

FIGURE 1.14 The scaling of the CGP and the M1 pitch in the past 3 decades and the projection into future nodes. The inset shows the layout of a FinFET-based logic standard cell.

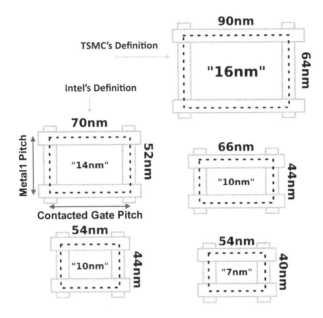

FIGURE 1.15 Simplified layout of a transistor showing the CGP and the M1 pitch for different technology nodes by different manufacturers.

(CMOS) logic standard cell design that includes both n-type and p-type transistors (NMOS and PMOS). To account for this, the integration density metric needs to consider the difference in the number of M1 tracks between logic processes. The recent trend is reducing the number of fins and M1 tracks but increasing the fin height to guarantee the drive current density. Figure 1.16 shows the scaling trend of the logic standard cell layouts in recent and projected generations. It is seen that the

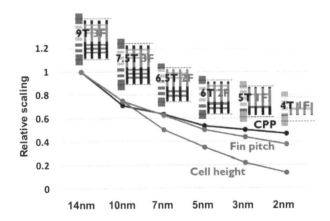

FIGURE 1.16 The scaling trend of the logic standard cell layouts in recent and projected generations. T is the number of M1 tracks and F is the number of fins.

Semiconductor Memory Technologies Overview

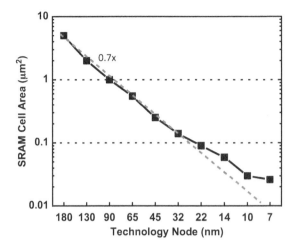

FIGURE 1.17 The scaling trend of high-density SRAM cell area in µm².

CGP/M1 pitch scaling slows down in leading-edge nodes; however, the cell height could scale more aggressively. For example, in the 5 nm node, there are only 2 fins for NMOS and 2 fins for PMOS, and 6 M1 tracks including top/bottom power lines. For future nodes at 3 nm and beyond, the number of stacked nanosheets could also be increased to further reduce the number of M1 tracks.

It is worth noting that the CGP/M1 metric is inaccurate in quantifying the density of the SRAM cells. SRAM cells rely largely on four metal wires: ground, power, WL, and BL, and foundries have optimized their layout rule that is much more compact than the logic layout rule even at the same-technology node. In practice is the absolute area in the unit of µm² is a direct measure for the SRAM cell density. Figure 1.17 shows the scaling trend of the high-density SRAM cell area in µm². Apparently, the SRAM scaling has deviated from the 0.7× scaling trend in the recent nodes, indicating a slowdown of Moore's law.

1.5 LOGIC TRANSISTOR TECHNOLOGY EVOLUTION

The semiconductor industry has made tremendous progress in the past few decades to sustain the continued technology scaling beyond simple geometrical reduction. The employed CMOS transistor optimization strategies considered several aspects: (1) increasing the on-state current (thus resulting in less delay and faster circuit operations); (2) suppressing the off-state current (thus minimizing the standby leakage power); (3) maintaining the gate-to-channel coupling and suppressing the drain-to-channel coupling, because the transistor is essentially a gate-controlled switch. To realize these goals, a few important technological innovations have been introduced to the logic transistor's materials and device structures, as shown in Figure 1.18.

First, the strained silicon technology was introduced in the 90 nm node [9]. The raised source/drain contacts used SiGe material instead of pure silicon, originally as

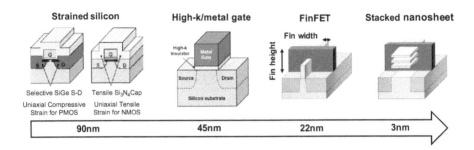

FIGURE 1.18 Important technological innovations to the logic transistor's materials and device structures at 90 nm, 45 nm, 22 nm, and possibly 3 nm.

a mechanism to reduce the series resistance. It turned out that the SiGe imposed a compressive strain on the PMOS transistor, and the strained silicon's crystal structure effectively changed the energy band structure, leading to an improvement in hole mobility. On the other hand, the Si_3N_4 capping layer induced tensile strain on the NMOS transistor. Strain engineering has been applied to all the technology nodes below 90 nm to improve carrier mobility.

Second, the high-k/metal gate technology was introduced in the 45 nm node [10]. To increase the effective gate capacitance and the gate-to-channel coupling, the thickness of the gate oxide needs to be shrunk. However, when SiO_2 approaches the sub-2 nm regime, the direct tunneling current exponentially increases due to the quantum mechanics, resulting in a substantial increase of the gate leakage. To increase the gate capacitance without further increasing the gate leakage, one natural way is to adopt a high-k dielectric of which the relative permittivity or the dielectric constant k value is higher than that of the SiO_2 ($\underline{k}$ = 3.9). The industry selected the HfO_2-based high-k dielectric (k = 20–25) to achieve the equivalent oxide thickness (EOT) below 1 nm when normalized to the SiO_2 while maintaining a physical thickness still larger than 3 nm. To mitigate the undesired effects brought by a high-k dielectric such as reduced carrier mobility due to the remote phonon scattering, the industry also made the decision to switch the heavily doped poly-silicon gate to the metal gate. To tune the threshold voltage, appropriate metal work function engineering is further required in all technology nodes below 45 nm.

Third, the FinFET technology was introduced at the 22 nm node [11]. This is a groundbreaking transition from the traditional planar structure to the non-planar structure. The main motivation is to combat the short channel effects when the gate length is reduced to the sub-50 nm regime such as a degraded sub-threshold slope (thus substantially increased off-state current). When the drain is so close to the source, the increase of drain voltage will significantly lower the energy barrier between the source and the channel, namely the drain-induced-barrier-lowering (DIBL) effect. As a result, the threshold voltage will be strongly influenced by the drain voltage. In other words, the gate tends to lose control of the carrier inversion in the channel; instead, the drain tends to gain control of the carrier injection to the channel. To improve the gate-to-channel coupling electrostatically, building a non-planar fin structure to allow extensive gate overlapping with three surfaces of the channel is helpful. Another advantage of FinFET is its improvement of the current density per physical footprint. Typically, the fin is very tall, and the fin height is becoming larger than the lateral fin pitch. Therefore, at the same physical footprint, FinFET could provide an improved on-state current. Owing to these attractive features, FinFET with tri-gate structure has become the mainstream technology and has enabled several generations of technology nodes from 22 nm node to 5 nm node.

Last but not the least, the stacked nanosheet transistor is expected to be introduced at the 3 nm node or beyond [12]. The ultimate structure that offers the best electrostatic gate-to-channel coupling is the gate-all-around (GAA) geometry. Combing the GAA with the tall fin structure gives an opportunity for the stacked nanosheet structure, which could further increase the on-current density per footprint if multiple layers of the nanosheet are stacked vertically. A more advanced design is the complementary n-type and p-type transistors (NMOS and PMOS) could be stacked on the same fin [13], realizing the monolithic 3D CMOS logic circuit modules. There is still plenty of room for device-level engineering and optimization for providing better performance and higher integration density in the coming years of the 2020s. These technological advances made in the logic process could directly benefit the SRAM design, and they could be potentially transferred to the peripheral circuit design of DRAM or NAND Flash that generally lags behind a few generations in its periphery.

NOTES

1. Prefixes for the scale used in this book: a, atto→10^{-18}; f, femto→10^{-15}; p, pico→10^{-12}; n, nano→10^{-9}; μ, micro→10^{-6}; m, milli→10^{-3}; k, kilo→10^{3}; M, mega→10^{6}; G, giga→10^{9}; T, tera→10^{12}; P, peta→10^{15}; E, exa→10^{18}; Z, zetta→10^{21}.
2. Sometimes, memory and storage can be used interchangeably.
3. B, Byte; b, bit; 1B = 8b.
4. In this book, without specification, DRAM is referred to the standalone DRAM, and eDRAM is used for the embedded DRAM.
5. Approximately about half of the revenue of a processor is allocated to the embedded memories.
6. A notable difference between a microprocessor and a microcontroller is that microcontroller is more application-specific and generally includes embedded NVMs for code storage.
7. The 3D stacked version of such a 2D matrix may be used for higher density design.
8. If $n = 2$, it means it stores digital information 00, 01, 10, and 11, and it is typically referred to as MLC. If $n = 3$, it is also referred to as triple-level cell (TLC). If $n = 4$, it is also referred to as quadruple-level cell (QLC).
9. Unit capacitance, unit delay, and unit area in the driver design is normalized to the load capacitance, delay, and area of a minimum-sized inverter at certain technology node, respectively.
10. Resistive memories include RRAM, PCM, and MRAM, which are discussed in more detail in Chapter 5.
11. CGP is also referred to CPP, contacted-poly-pitch for a historical reason as the gate used to be made of poly-silicon before the high-k/metal gate was adopted. CGP is also the same as the center-to-center distance from the source contact via to the drain contact via.

REFERENCES

[1] Zetta-scale data, https://www.forbes.com/sites/tomcoughlin/2018/11/27/175-zettabytes-by-2025/

[2] C. Gonzalez, E. Fluhr, D. Dreps, D. Hogenmiller, R. Rao, J. Paredes, M. Floyd, et al., "POWER9™: a processor family optimized for cognitive computing with 25Gb/s accelerator links and 16Gb/s PCIe Gen4," *IEEE International Solid-State Circuits Conference (ISSCC)*, 2017, pp. 50–51, doi: 10.1109/ISSCC.2017.7870255.

[3] Standalone memory market size in 2020, https://www.icinsights.com/data/articles/documents/1375.pdf

[4] 3D X-point technology, https://www.intel.com/content/www/us/en/architecture-and-technology/intel-micron-3d-xpoint-webcast.html

[5] M.-F. Chang, A. Lee, P.-C. Chen, C.J. Lin, Y.-C. King, S.-S. Sheu, T.-K. Ku, "Challenges and circuit techniques for energy-efficient on-chip nonvolatile memory using memristive devices," *IEEE Journal on Emerging and Selected Topics in Circuits and Systems*, vol. 5, no. 2, pp. 183–193, June 2015, doi: 10.1109/JETCAS.2015.2426531.

[6] G.E. Moore, "Cramming more components onto integrated circuits," *Proceedings of the IEEE*, vol. 86, no. 1, pp. 82–85, January 1998, doi: 10.1109/JPROC.1998.658762.

[7] The Moore's Law slowing down, https://semiengineering.com/the-impact-of-moores-law-ending/

[8] H.-S.P. Wong, K. Akarvardar, D. Antoniadis, J. Bokor, C. Hu, T.-J. King-Liu, S. Mitra, J.D. Plummer, S. Salahuddin, "A density metric for semiconductor technology [point of view]," *Proceedings of the IEEE*, vol. 108, no. 4, pp. 478–482, April 2020, doi: 10.1109/JPROC.2020.2981715.

[9] T. Ghani, M. Armstrong, C. Auth, M. Bost, P. Charvat, G. Glass, T. Hoffmann, et al., "A 90nm high volume manufacturing logic technology featuring novel 45nm gate length strained silicon CMOS transistors," *IEEE International Electron Devices Meeting (IEDM)*, 2003, pp. 11.6.1–11.6.3, doi: 10.1109/IEDM.2003.1269442.

[10] K. Mistry, C. Allen, C. Auth, B. Beattie, D. Bergstrom, M. Bost, M. Brazier, et al., "A 45nm logic technology with high-k+metal gate transistors, strained silicon, 9 Cu interconnect layers, 193nm dry patterning, and 100% Pb-free packaging," *IEEE International Electron Devices Meeting (IEDM)*, 2007, pp. 247–250, doi: 10.1109/IEDM.2007.4418914.

[11] C. Auth, C. Allen, A. Blattner, D. Bergstrom, M. Brazier, M. Bost, M. Buehler, et al., "A 22nm high performance and low-power CMOS technology featuring fully-depleted tri-gate transistors, self-aligned contacts and high density MIM capacitors," *IEEE Symposium on VLSI Technology*, 2012, pp. 131–132, doi: 10.1109/VLSIT.2012.6242496.

[12] N. Loubet, T. Hook, P. Montanini, C.-W. Yeung, S. Kanakasabapathy, M. Guillom, T. Yamashita, et al., "Stacked nanosheet gate-all-around transistor to enable scaling beyond FinFET," *IEEE Symposium on VLSI Technology*, 2017, pp. T230–T231, doi: 10.23919/VLSIT.2017.7998183.

[13] C.-Y. Huang, G. Dewey, E. Mannebach, A. Phan, P. Morrow, W. Rachmady, I.-C. Tung et al., "3-D self-aligned stacked NMOS-on-PMOS nanoribbon transistors for continued Moore's Law scaling," *IEEE International Electron Devices Meeting (IEDM)*, 2020, pp. 20.6.1–20.6.4, doi: 10.1109/IEDM13553.2020.9372066.

2 Static Random Access Memory (SRAM)

2.1 6T SRAM CELL OPERATION

2.1.1 SRAM Array and 6T Cell

Static random access memory (SRAM) is the mainstream embedded memory technology that is primarily used as an on-chip cache for processors. "Static" means that the data are maintained as long as the power supply is on. "Random access" means that each bit of the data could be read/written independently. The SRAM sub-array is usually organized as shown in Figure 2.1(a). Besides the common peripheral circuits such as decoder, MUX, sense amplifier, and write driver, an additional pre-charge and equalizer module is used. This is to serve the complementary bit line nature of the SRAM array, which has a bit line (BL) and a bit line_bar ($\overline{BL}$). Figure 2.1(b) shows the detailed circuit schematic for the column-side periphery. The read/write selection MUX decides the operation modes between the read and the write. During the read, the sense amplifier is enabled to sense the small signal difference between BL and $\overline{BL}$, then amplify the signal to digital "0" and "1" and the data are then latched in the output flip-flop. During the write, the write driver is activated, and the to-be-written data and its complementary data are generated and delivered to the cell via BL and $\overline{BL}$.

Figure 2.2(a) shows the circuit schematic of the commonly used 6-transistor (6T) SRAM cell.[1] The core of the 6T cell includes two cross-coupled inverters (INV1 and INV2). N1 and N2 nodes are called the storage nodes of the SRAM cell, as the data are essentially represented as charges that are stored at one of these two nodes. Due to the symmetric nature of the 6T cell, the data pattern is always complementary between N1 and N2. If N1 stores "1" with a voltage at the power supply (V_{DD}), then N2 must store "0" with a voltage at the ground. Therefore, only 1-bit information is stored in the 6T cell, and it is typically regarded as being stored at N1. The storage nodes are connected to BL and $\overline{BL}$ through two access transistors, whose gates are controlled by the same wordline (WL). Figure 2.2(b) shows the transistor-level schematic of the 6T cell, with four NMOS transistors, and two PMOS transistors. The six transistors are categorized into three groups: pull-down (PD) NMOS, pull-up (PU) PMOS, and pass-gate (PG) NMOS.

2.1.2 Principles of Hold, Read and Write

The hold operation is to maintain the data if no read or write operation is being performed. In the hold mode, WL is grounded; thus, both PG transistors are turned off. As a result, N1 and N2 nodes are isolated from BL and $\overline{BL}$ even if BL and $\overline{BL}$

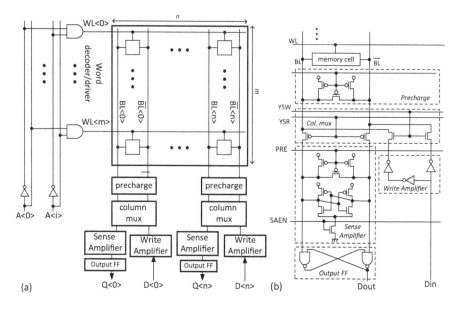

FIGURE 2.1 (a) SRAM sub-array diagram with peripheral circuits. (b) Circuit schematic for the column-side periphery.

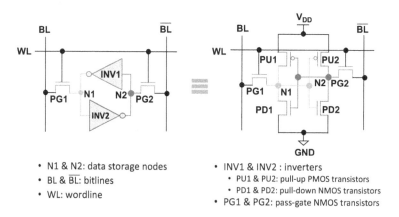

- N1 & N2: data storage nodes
- BL & $\overline{BL}$: bitlines
- WL: wordline

- INV1 & INV2 : inverters
- PU1 & PU2: pull-up PMOS transistors
- PD1 & PD2: pull-down NMOS transistors
- PG1 & PG2: pass-gate NMOS transistors

FIGURE 2.2 Circuit schematic of the 6-transistor (6T) SRAM cell.

are both biased at V_{DD}. The positive feedback of the cross-coupled inverters helps to maintain the data as long as the power supply is on.

The SRAM read operation is shown in Figure 2.3(a). In the following discussion, "0" is assumed to be stored at N1 and "1" is assumed to be stored at N2. Before the read operation, both BL and $\overline{BL}$ are pre-charged to V_{DD} by the 3-transistor-based equalizer circuit (as shown in Figure 2.1(b)). Here, BL and $\overline{BL}$

Static Random Access Memory (SRAM)

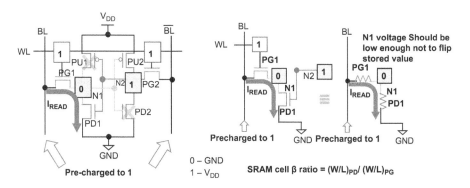

FIGURE 2.3 (a) Principle of the SRAM read operation with major current flow path. (b) Approximation of discharge current path as resistor-based voltage divider during the read.

could be modeled as two capacitors as long interconnect wires introduce parasitic capacitance, thus holding the charges when being pre-charged to V_{DD}. During the read, the equalizer is disabled, and WL is activated by a voltage pulse (i.e., with V_{DD}), now both PG transistors are turned on. The left branch of the 6T cell will see a discharge current from BL to ground through PG1 and PD1 because the N1 stores "0" and BL/N1 becomes the drain/source of the PG1 (an NMOS transistor). Therefore, current flows from BL to N1, and this current must be sunk to the ground through PD1. Hence, this current will slightly raise up the N1 voltage as N1 becomes the drain of the PD1. Due to this discharge current from BL, the BL voltage will decay from V_{DD} over time. The right branch of the 6T cell will not see any significant discharge current from $\overline{BL}$, because both N2 and $\overline{BL}$ should stay at V_{DD} and an equal potential of PG2's source/drain should not flow any current. Now, there is a voltage difference being developed between BL and $\overline{BL}$, namely the sense margin ΔV, and BL voltage is slightly smaller than $\overline{BL}$ voltage in this example of reading "0". After a certain signal development time, the sense amplifier is enabled, and such a small voltage difference will be amplified due to its differential input nature. One representative example of the latch-based sense amplifier design is shown in Figure 2.1(b). Initially, the SAEN signal is off, and both sides of the latch in the sense amplifier are pre-charged to V_{DD} by the equalizer. When the SAEN signal is on, the small voltage difference transferred from BL and $\overline{BL}$ will flip the state of the latch. In the case of reading "0" from N1, a smaller BL voltage will push down the left side of the latch to the ground, and V_{DD} remains at the right side of the latch. As a result, such a data pattern is latched into the output flip-flop, and data of the DOUT node are the same as the stored "0" data in this 6T cell.

One critical design consideration for read is to ensure that the stored "0" will not be accidentally flipped by the discharge current. As aforementioned, the N1 node voltage is raised during the read operation, resulting in a risk to flip N1 to "1" if the N1 node voltage is too high to trigger the flipping of the cross-coupled inverters (especially when the noise exists). Hence, there is a limitation on the maximum voltage allowed at N1. During the read, only two transistors are active and involved (PG1 and PD1). To analyze the requirements of a safe read condition, PG1/PD1 could be approximated as simple resistors. Therefore, the discharge current path becomes a two-resistor-based voltage divider, as shown in Figure 2.3(b). To ensure that the N1 node voltage is not being substantially raised, the resistance of PD1 should be smaller than that of PG1. In other words, the conductance of PD1 should be larger than that of PG1. This will be translated to a larger current drivability for PD1 than that for PG1. Since both PD1 and PG1 are NMOS transistors, their width/length (W/L) ratio should be sized appropriately and PD1 should be wider than PG1, in principle, indicating the PD transistor to be stronger than the PG transistor. The SRAM cell beta ratio β is defined as $(W/L)_{PD}/(W/L)_{PG}$, and it is a key parameter that determines the safe read condition. Typically, $\beta = 2$ is used in most mature technology nodes in the planar transistor era.

Figure 2.4 shows the timing diagram and waveforms for the read operation. First, BL and $\overline{BL}$ are pre-charged to V_{DD} and then WL is turned on. It is seen that not only does the N1 node voltage increases slightly, but also the N2 node voltage decreases slightly due to the positive feedback nature of the cross-couple inverters. Despite such disturbances, N1 and N2 voltages should be well separated to avoid accidental flipping during the read. After a sufficient voltage sense margin (defined as the difference between BL and $\overline{BL}$ voltages) is developed, the SAEN signal is turned on. Due to the latching effect by the sense amplifier, the N1 node is recovered to "0" and the N2 node is recovered to "1". After WL is off, the equalizer is turned on to pre-charge both BL and $\overline{BL}$ to V_{DD}. The read speed is primarily determined by the PG transistor's drivability. For example, if BL parasitic capacitance $C_{BL} = 50$ fF, to achieve a

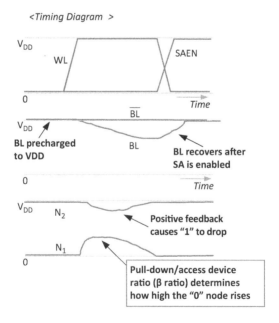

FIGURE 2.4 Timing diagram and waveforms for the SRAM read operation.

Static Random Access Memory (SRAM)

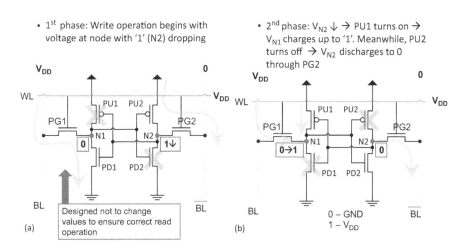

FIGURE 2.5 (a) Principle of the SRAM write operation in the 1st phase that "1" to "0" transition is initiated. (b) "0" to "1" transition is followed in the 2nd phase.

sense margin $\Delta V = 100$ mV within a delay $\Delta t = 0.5$ ns, the PG transistor's current could be approximated as $I_{PG} = C_{BL} \times \Delta V/\Delta t = 10$ µA.

The SRAM write operation is shown in Figure 2.5. In the following discussion, "0" is assumed to be stored at N1 and "1" is assumed to be stored at N2. The write operation tends to flip the state by storing "1" to N1 and "0" to N2. To prepare for the write, BL is biased at V_{DD} and $\overline{BL}$ is biased at the ground by the write driver given the to-be-written data from the external data bus. During the write, WL is activated by a voltage pulse (i.e., with V_{DD}), and both PG transistors are turned on. The first phase of the write operation is shown in Figure 2.5(a). The left branch of the 6T cell will see the same bias condition as the read operation where N1 is supposed to be slightly raised by the current flowing through PG1 and PD1. However, the N1 node will not trigger the flip from "0" to "1" as ensured by the safe read condition. The right branch of the 6T cell will see a current flow through PU2 and PG2 from V_{DD} to $\overline{BL}$ which is grounded. Therefore, the N2 node will decay its voltage due to this discharge current. The second phase of the write operation is shown in Figure 2.5(b). Note that the N2 node is also the gate of PU1. When the N2 voltage is sufficiently low, a negative gate-to-source voltage is applied to the PU1 (a PMOS transistor); thus, PU1 is turned on. Then PU1 could pass additional current from V_{DD} to further charge up the N1 node. The N1 node voltage will increase toward V_{DD}; meanwhile, PU2 will be turned off as the N1 node is the gate of PU2. This is a positive feedback loop, and eventually, N1 will reach V_{DD}, and N2 will decay to the ground. The write operation is now complete.

To summarize, the write operation is initiated by the "1" to "0" transition, as the "0" to "1" transition is prohibited by the safe read condition. Since the write operation is initially involved with only two transistors, PU2/PG2 could be approximated as simple resistors to analyze the requirements of a safe write condition. Similarly, the discharge current path becomes a two-resistor-based voltage divider as shown in Figure 2.6. To ensure the N2 node voltage being substantially lowered down (to activate PU1), the resistance of PU2 should be larger than that of PG2. In other words,

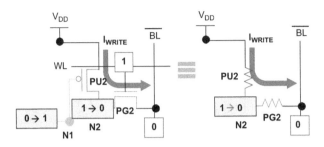

FIGURE 2.6 Approximation of discharge current path as resistor-based voltage divider during the write.

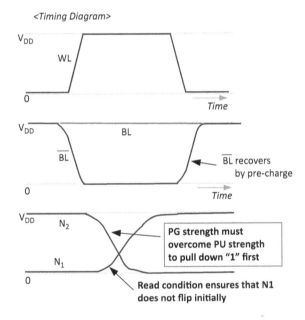

FIGURE 2.7 Timing diagram and waveforms for the SRAM write operation.

the conductance of PU2 should be smaller than that of PG2. This will be translated to a larger current drivability for PU2 than that for PG2. Note that PU2 is a PMOS transistor and PG2 is an NMOS transistor. Even with the same transistor width, NMOS typically has larger drivability than PMOS because electrons have higher mobility than holes,[2] indicating PG transistor being weaker than PU transistor. The SRAM cell gamma ratio γ is defined as $(W/L)_{PU}/(W/L)_{PG}$, and it is a key parameter that determines the safe write condition. Typically, $\gamma = 1$ is used in most mature technology nodes in the planar transistor era.

Figure 2.7 shows the timing diagram and waveforms for the write operation. First, BL and $\overline{BL}$ are biased to V_{DD} and ground, then WL is turned on. It is seen that the N2 node decays more rapidly than the N1 node ramps. This agrees with the earlier discussion that the "1" to "0" transition occurs first, then the "0" to "1" transition follows. Eventually, the crossover of the N2 and N1 voltages is triggered and the flipping is then complete. After WL is off, the equalizer is turned on to pre-charge both BL and $\overline{\text{BL}}$ to V_{DD}.

2.2 SRAM STABILITY ANALYSIS

2.2.1 Static Noise Margin

The SRAM cell stability could be characterized by static and dynamic analyses [1]. The static analysis assumes that the stimulus that tends to flip the cell (either by the noise or by the write operation) is present forever, which of course is not a realistic assumption. Nevertheless, it is easier to start with the static analysis, which could give intuitive design guidelines. The data stored in the SRAM is resilient to a certain degree of noise in the circuits thanks to the positive feedback nature offered by the cross-coupled inverters. The static noise margin (SNM) is widely used to measure the stability against the disturbance caused by the noise. The SNM could be classified into three types: Hold SNM (H-SNM), Read SNM (R-SNM), and Write SNM (W-SNM), which will be discussed as follows.

The hold operation has the PG transistors off; thus, the cross-coupled inverters are isolated from BL and $\overline{\text{BL}}$. The H-SNM is essentially characterized by the voltage transfer curve (VTC) of these two cross-coupled inverters. Figure 2.8 shows the butterfly curve consisting of two VTCs. Here, the x-axis is the N1 voltage, and the y-axis is the N2 voltage. For INV2, N1 is the input and N2 is the output. As the N1 voltage is swept from ground to V_{DD}, the N2 voltage decays from V_{DD} to ground following the red VTC. For INV1, N2 is the input and N1 is the output. Therefore, the x-axis and the y-axis should exchange their roles. As the N2 voltage is swept from ground to V_{DD}, the

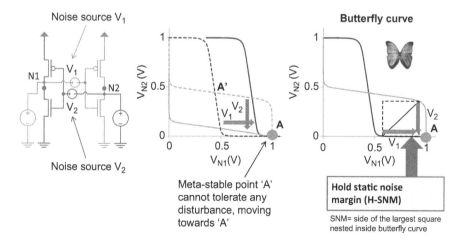

FIGURE 2.8 The butterfly curve consisting of two voltage transfer curves of the cross-coupled inverters. The hold static noise margin (H-SNM) is shown.

N1 voltage decays from V_{DD} to ground following the blue VTC. There are three intersection points between these two VTCs. Two of them at the corners are the stable points that store the memory states, while the middle one is the meta-stable point that could be easily disturbed by noise and flipped into either of the two stable points. The static noise is modeled as a voltage source that is applied to one of the storage nodes, which effectively shifts one of the VTC so that three stable points become two (with only one stable point). Now, the cell is deterministically flipped to the only stable point A in this example. The maximum noise that the cell can tolerate during the hold is measured by the maximum nested square that can fit into the two VTCs. By definition, H-SNM is the side length of this nested square. In theory, the maximum nested square could be found by rotating the x-axis and the y-axis by 45° to calculate the maximum distance between the two VTCs. In an ideal case, the two VTCs are symmetric along the straight line ($V_{N2} = V_{N1}$) that follows along the 45° angle in this plot.

The read operation turns the PG transistors on; thus, the cross-coupled inverters are connected to BL and $\overline{BL}$. To help to analyze the distorted VTC (red one for INV2) under the read condition, one could consider the three transistors of the right branch (PD2, PU2, and PG2) as shown in Figure 2.9. $\overline{BL}$ is pre-charged to V_{DD} and in the static analysis, it could be assumed to be biased at V_{DD}. Again, the x-axis is the N1 voltage, and the y-axis is the N2 voltage. As the N1 voltage is swept from the ground to V_{DD}, the N2 voltage decays from V_{DD} to a certain low level. Different from the hold operation, this low level of the N2 voltage is non-zero, but is above zero, which is essentially determined by the voltage divider effect between PG2 and PD2. The larger the β ratio, the smaller the low level. As discussed earlier, the voltage increase to this low level needs to be minimized during the read to avoid accidental flipping. Similarly, the other distorted VTC (blue one for INV1) could be determined this way or simply by mirroring the VTC (red one for INV2) along the 45° line. As a result, the butterfly curve now has a smaller window to fit in the nested square than the hold case. The maximum noise that the cell can tolerate during the read is measured by the maximum nested square that can fit into the two distorted VTCs. By

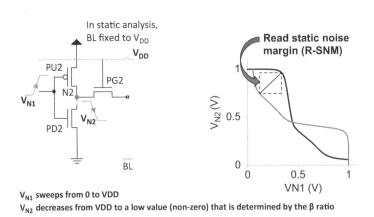

FIGURE 2.9 The butterfly curve consisting of two distorted voltage transfer curves during the read. The read static noise margin is shown.

Static Random Access Memory (SRAM)

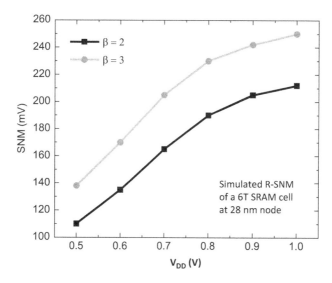

FIGURE 2.10 Read static noise margin as a function of power supply V_{DD} and cell β ratio for a 28 nm node 6T SRAM cell.

definition, R-SNM is the side length of this nested square. It is clearly seen that R-SNM is always smaller than H-SNM. In other words, the read operation is more vulnerable to the noise than the hold operation. This is because during the read, the storage node that stores "0" is slightly raised thus it is easier to flip to "1". Figure 2.10 shows the simulated R-SNM as a function of V_{DD} and β ratio in a 28 nm 6T SRAM cell. It is seen that R-SNM decreases with lowering V_{DD}, and increasing β ratio will boost R-SNM at the expense of a larger area. For a given technology node, V_{DD} thus has a minimum level that ensures a reasonable SNM (e.g., >100 mV).

The write operation has asymmetric VTCs between the left branch and the right branch as shown in Figure 2.11. Assumedly that N2 initially stores "0" and it will be written to "1",[3] the right branch has the same bias condition as the read operation, which generates the (red) VTC similarly as in Figure 2.9. What differs from the read is the left branch, where BL is now biased at ground. As the N2 voltage is swept from ground to V_{DD}, the N1 voltage decays from a certain low level to ground, leading to the other (blue) VTC for the left branch. This certain low level is essentially determined by the voltage divider effect between PU1 and PG1. The smaller the γ ratio, the smaller the low level. Since there is only one intersection point between the two VTCs, a deterministic write operation will occur that flips the state as N1 stores "0" and N2 stores "1". To avoid uncertainty during the write, the two VTCs should be separated as far as possible without other intersection points. Therefore, the maximum noise that the cell can tolerate during the write is measured by the minimum nested square that can fit into the two VTCs. By definition, W-SNM is the side length of this nested square.

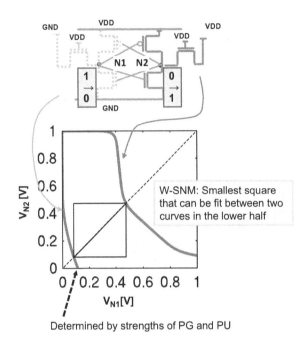

FIGURE 2.11 Asymmetric voltage transfer curves between the left branch and the right branch during the write. The write static noise margin is shown.

2.2.2 N-curve

Although the R-SNM metric is a common measure for the read voltage noise margin in the static analysis, it has several drawbacks. For example, it is not straightforward to derive the R-SNM as one needs to perform post-processing of the data (i.e., rotating the x-axis and the y-axis by 45° and finding the nested square). To simplify the testing procedure and make it compatible with the inline tester, the N-curve method is proposed [2]. The N-curve method just requires probing the storage node of the SRAM cell by a voltage source, both static voltage noise margin (SVNM) and static current noise margin (SINM) are extracted without post-processing, as shown in Figure 2.12. It is assumed that N1 stores "1" and N2 stores "0", BL is biased at V_{DD} and an external voltage source is attached to N2. As the input voltage (V_{in}) from the external source increases from ground to V_{DD}, the current that flows through the voltage source is metered. Initially, when V_{in} is low, the current flow is from the BL to N2 which will be split between PD2 and the voltage source. The current I_{in} is thus negative (sinking to the voltage source). As the N2 voltage is swept up, more current will flow through PD2, and at certain point B, the current flow to the voltage source becomes zero. Then the N2 voltage is further forced to increase, the voltage source will inject current to the N2 node; thus, I_{in} becomes positive. At a certain point P, the N2 voltage becomes so large that it triggers the flipping; thus, the N-curve snaps back. PD2 is turned off (as its gate voltage N1 node becomes low), but PU2 is turned on, contributing additional current to the N2 node, which can only sink to the voltage

Static Random Access Memory (SRAM)

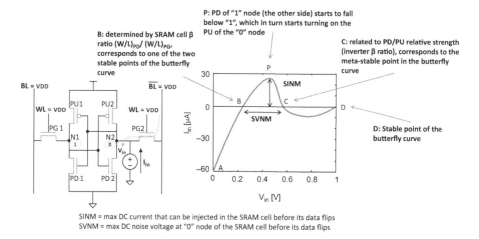

FIGURE 2.12 The N-curve testing protocol of the SRAM cell. The static voltage noise margin (SVNM) and the static current noise margin (SINM) are shown.

source (thus I_{in} becomes negative again). Finally, when V_{in} reaches V_{DD}, the SRAM cell reaches a stable state and I_{in} becomes zero. The SVNM is defined as the voltage distance between points B and C, and SINM is defined as the current peak at point P.

Figure 2.13 shows the traces of the entire process of the N-curve testing using the output characteristics (drain current vs. gate voltage) of the three types of transistors (PD, PU, PG). The sum current from the three transistors toward the N2 node

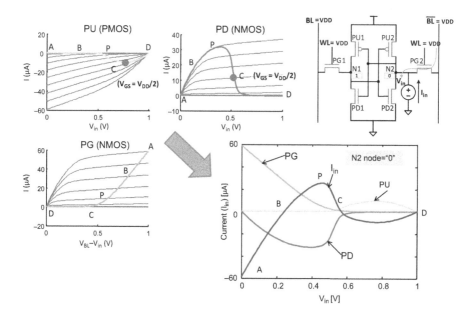

FIGURE 2.13 The traces of the entire process of the N-curve testing using the output characteristics (drain current vs. gate voltage) of the three types of transistors (PD, PU, PG).

- Butterfly curve
  - $\Sigma (I_{PG}+I_{PU}+I_{PD}) = 0$ always holds
  - SNM is defined by noise <u>between the two storage nodes</u>
  - More relevant for transistor mismatch
- N curve
  - $\Sigma (I_{PG}+I_{PU}+I_{PD})=0$ for V_{in} node only when $I_{in} =0$ (Points A, B and C)
  - <u>Noise between "0" node and GND</u>
  - More relevant for noise current injection

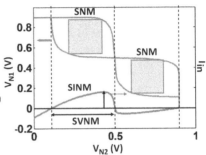

FIGURE 2.14 Correlation between the butterfly curve and the N-curve.

determines the I_{in} following the Kirchhoff law. Figure 2.14 shows the correlation between the butterfly curve and the N-curve. Both curves belong to the static analysis. In fact, the critical points between the two curves are correlated. The butterfly curve having $\Sigma(I_{PG} + I_{PU} + I_{PD}) = 0$ always holds, and R-SNM is defined by the noise between the two storage nodes; thus, it is more relevant for the transistor mismatch due to process variation. The N-curve will have $\Sigma(I_{PG} + I_{PU} + I_{PD}) = 0$ for V_{in} node only when $I_{in} = 0$ (Points B, C, and D) and it essentially captures the noise between the N2 node and ground; thus, it is more relevant for the noise current injection. The advantage of the N-curve is that the SVNM and SINM could be read out directly from the plot without further post-processing. It should be noted that SVNM defined in the N-curve is larger than the R-SNM defined in the butterfly curve.

2.2.3 Dynamic Noise Margin

The dynamic analysis is more relevant to the practical scenarios of SRAM operations, as the read/write is performed in the pulse mode or the noise is usually for a short period of time. For the hold mode, the noise could be modeled as a current source injection that lasts for a certain amount of time. Assumed that N1 stores "1", N2 stores "0" and the noise current is injected to N2, the waveform of N1 and N2 voltages could be simulated as a function of time. The time-domain waveform could be mapped to the N1 voltage versus the N2 voltage diagram as a trajectory. As the noise current injects charges to the N2 node, N2 voltage increases above ground, and N1 voltage may drop. If the noise duration is short (e.g., disappearing after 200 ps in this example), N2 and N1 voltages will return to ground and V_{DD}, respectively, as shown in Figure 2.15(a). This automatic recovery of the memory state is owing to the positive feedback nature of the cross-coupled inverters. However, if the noise duration is long enough that the N2 voltage rises beyond the flipping point, then the recovery fails and the state will flip, as shown in the time-domain waveform and the trajectory in the N1 voltage versus the N2 voltage diagram in Figure 2.15(b). Essentially, there is a critical charge amount (Q_{crit}) that will trigger the flipping, which depends on the product of noise current amplitude and its duration. As the technology scales down, the parasitic capacitances of the

Static Random Access Memory (SRAM)

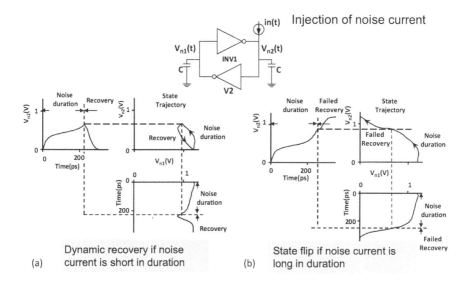

FIGURE 2.15 Dynamic analysis of the transient noise (modeled as a current injection to the storage node "0"). The time-domain waveform of N1 and N2 voltages and their trajectory in N1 voltage vs. N2 voltage diagram. (a) Dynamic recovery if noise is short; (b) State flip if the noise is long.

storage node that are contributed by the three transistors (e.g., drain to body capacitances of PD, PG, and PU) may reduce; thus, Q_{crit} may shrink, making the SRAM cell more vulnerable.

In the N1 voltage versus the N2 voltage diagram, one can define the concept of "separatrix" [3], which defines the boundary of stability of an SRAM cell when the pass gate transistor (PG) is off. On either side of the separatrix, the SRAM cell's two storage nodes voltage will be attracted toward the two stable points at the upper left and lower right of the diagram. When the cross-coupled inverters are perfectly matched, the separatrix is a straight line with a 45° angle in the plot. When the cross-coupled inverters are not perfectly matched, the separatrix will be a curve dividing the diagram. With a certain-level current injection to the storage node, one can define a critical time (T_{across}) that the node voltage's trajectory requires to cross the separatrix. Figure 2.16 shows the separatrix diagram and the conflict between the read and write operation. During the read, the node that stores "0" is raised by the PG transistor current, which is similar to a noise current injection as discussed earlier in the hold mode. Therefore, the read pulse width (T_R) needs to be sufficiently short to avoid the accidental write. In the separatrix diagram, $T_R < T_{across}$ should be satisfied. During the write, the node that stores "1" is pulled down by the PG transistor current, which will take some time to discharge before reaching the flipping point. Therefore, the write pulse width (T_W) needs to be sufficiently long to ensure a successful write. In the separatrix diagram, $T_W > T_{across}$ should be satisfied. Fundamentally, the conflict between the read and the write originates from the fact they share the same PG transistor in the 6T cell. From the stability point of view, a weaker PG transistor (compared to PD transistor) is preferred during the read to isolate the cell from external BL or $\overline{BL}$; from the

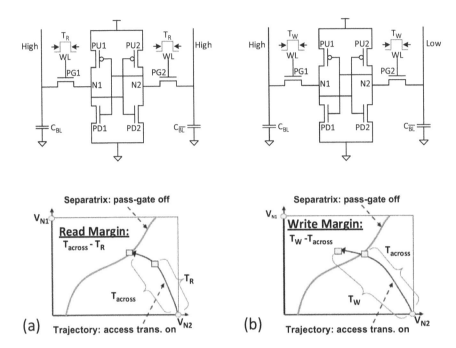

FIGURE 2.16 The separatrix diagram that shows the conflict between read and write operations. (a) The read pulse should be shorter than the critical time to cross the separatrix boundary. (b) The write pulse should be longer than the critical time to cross the separatrix boundary.

programming efficiency point of view, a stronger PG (compared to PU transistor) is preferred during the write to couple the cell to external BL or $\overline{BL}$.

Compared to the static analysis, the dynamic analysis is more accurate. Therefore, the estimated H-SNM or R-SNM is pessimistic for hold/read in practice as the noise duration or the read pulse is not infinite as assumed in static analysis. On the other hand, the estimated W-SNM is optimistic for write in practice as the write pulse is not infinite as assumed in static analysis. To characterize the SRAM's performance (for a given process technology), Shmoo plot is widely used as shown in Figure 2.17. The SRAM array is tested by varying the power supply V_{DD}, and the clock frequency (f for WL pulse) as a two-dimensional sweep, and the pass/fail is recorded for each combination of V_{DD} and f. Typically, failure occurs when the clock frequency is too high or when the V_{DD} is too low. When the clock frequency is too high, write failure dominates as there is not sufficient time to guarantee the state flipping. When the V_{DD} is too low, read failure dominates as there is no sufficient voltage headroom for the noise margin. Shmoo plot basically determines the envelope of SRAM's operating conditions. Generally, higher V_{DD} is beneficial to faster access.

2.2.4 Read/Write-Assist Schemes

To mitigate the conflict between the read and the write, read- and write-assist schemes could be applied. In general, there are two approaches to improve the

Static Random Access Memory (SRAM) 35

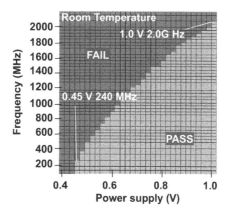

FIGURE 2.17 An example of the Shmoo plot of SRAM array level functionality (pass or fail) as a function of power supply V_{DD} and clock frequency f.

stability during the read and the efficiency during the write: 1) modifying the biasing and pulsing scheme; 2) adding additional transistors to the SRAM cell.

First, a dual-power supply could be used for a 6T cell, as shown in Figure 2.18(a). For the read operation, WL could be lowered than the V_{DD} of the cross-coupled inverters. As a result, the strength of the PG transistor is weakened, thus less-disturbed current flows into the storage node that stores "0". Therefore, the read stability is improved. The trade-off here will be a slower read speed. For the write operation, a negative BL pulse could be applied, facilitating the pull-down of the storage node that stores "1". Therefore, the write efficiency is improved. The overhead here will be the circuit design for generating negative voltage on-chip.

Second, if considering the entire SRAM array, there is a conflict between the selected columns and the unselected columns. During the write, the selected columns prefer a sufficiently long WL pulse to ensure a successful write; however, the unselected columns, as they share the same WL will suffer the same long WL pulse,

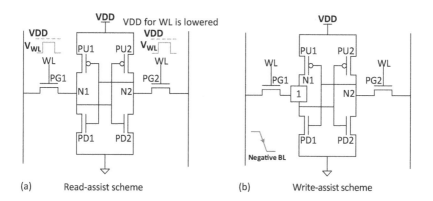

FIGURE 2.18 (a) The dual-power supply scheme to improve the read stability. (b) The negative BL scheme to improve the write efficiency.

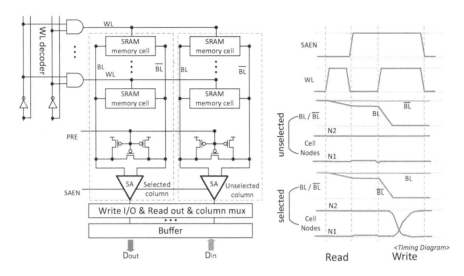

FIGURE 2.19 The read-modify-write (RMW) scheme with different WL pulse widths to resolve the different requirements for the read and the write in one SRAM array considering the selected and unselected columns.

and the cells that store "0" are susceptible to flip accidentally. To decouple the different requirements on the WL pulse width, a read-modify-write (RMW) scheme [4] is proposed in Figure 2.19. The idea is to split the write into two phases: First, all columns are read out using a short WL pulse (to avoid accidental write) and a sense amplifier is equipped per column. Next, the WL is turned on again for a long enough duration that provides sufficient write margin. In the second phase, cells in all the columns are written with either new data (for the selected columns) or old data (for the unselected columns). The sense amplifiers are used as write drivers to write the data back (for the unselected columns). The regular read operation is similar to the first phase of the RMW, but post-sense-amplifier multiplexing is used to select appropriate columns to transfer the data to the external data bus.

An orthogonal approach to modify the SRAM bit cell structure by adding more transistors. One commonly used bit cell is the read-decouple 8T cell [5], as shown in Figure 2.20. The write operation is still performed using the same scheme as for the previously discussed 6T cell. But the read path is now decoupled from the write path by two additional NMOS transistors (M7 and M8), read word line (RWL), and read bit line (RBL). The M7 transistor's gate is controlled by the storage node N1, and the M8 transistor's gate is controlled by RWL. Only when both RWL and N1 voltages are high, a discharge current exists and the RBL voltage will decay (to be sensed as "0"). Otherwise, the RBL voltage will remain (to be sensed as "1"). Such a read path is completely disturb-free to the storage nodes; thus, the read noise margin is now the same as the hold noise margin. To summarize, the read-decouple 8T cell could significantly improve the read stability, but at the expense of a noticeable area overhead.

Static Random Access Memory (SRAM)

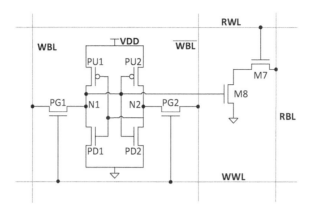

FIGURE 2.20 Circuit schematic of the read-decouple 8T SRAM cell that substantially improves read stability.

2.3 SRAM'S LEAKAGE

2.3.1 Transistor's Sub-threshold Current

To understand the SRAM's standby power consumption in the hold mode during idle or standby, it is useful to review the mechanism of the transistor's leakage mechanism using NMOS as an example. First, the transistor suffers from the sub-threshold current, where the current is mainly contributed by the carrier diffusion over the potential barrier between the source and the channel. In the transistor's transfer characteristics (log (I_D) vs. V_G), the drain current exponentially reduces when the gate voltage is below the threshold voltage (V_{th}) as shown in Figure 2.21(a). The off-state current (I_{off}) is defined as I_D when V_G is zero and V_D is V_{DD}, while the on-state current (I_{on}) is defined as I_D when both V_G and V_D are V_{DD}. In such a semi-log plot, a straight line in the sub-threshold regime indicates a sub-threshold slope (SS), which is defined as the following equation.

$$S = \left(\frac{d\log I_D}{dV_G}\right)^{-1} = \frac{\partial V_G}{\partial \Psi_S} \frac{\partial \Psi_S}{\partial \log I_D} = \left(1 + \frac{C_{dm}}{C_{ox}}\right) \frac{kT}{q} \ln(10) = m \times 2.3 \frac{kT}{q} \quad (2.1)$$

The SS could be modulated by two factors: the first term is the derivative of the gate voltage V_G with respect to the surface potential φ_s (reflecting the gate-to-channel coupling ratio), namely the body factor (m); the second term is the derivative of the surface potential to the drain current (reflecting the Boltzmann distribution that carriers are thermally excited over the potential barrier). The first term m represents the relative change of the surface potential with respect to the change of the gate voltage through a two-capacitor-based voltage divider model between the oxide capacitance C_{ox} and the silicon depletion capacitance C_{dm}, as shown in Figure 2.21(b). Note that $C_{ox} = \varepsilon_{ox}/t_{ox}$, and $C_{dm} = \varepsilon_{si}/W_{dm}$, and $\varepsilon_{si} \sim 3\varepsilon_{ox}$,[4] so the body factor m could be reformulated by the depletion width (W_{dm}) in silicon channel and the oxide thickness (t_{ox}) as follows.

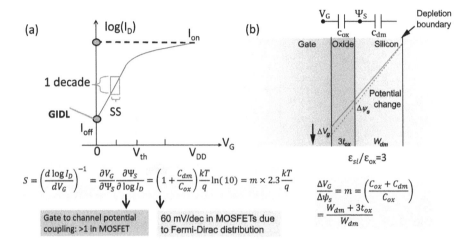

FIGURE 2.21(A) Transistor leakage mechanism showing the sub-threshold current and the gate-induced-drain-leakage (GIDL) current in log(I_D) vs. V_G transfer characteristics. (b) Schematic of the capacitor-based voltage divider model to illustrate the gate to channel coupling, namely the body factor (m).

$$m = \left(\frac{C_{ox} + C_{dm}}{C_{ox}}\right) = \frac{W_{dm} + 3t_{ox}}{W_{dm}} \quad (2.2)$$

The second term $2.3\ kT/q$ are determined by the temperature and the physical constant.[5] Since the body factor m is always larger than 1, the SS has a lower-bound limit of 60 mV/dec at room temperature. This suggests that to shrink the drain current by one decade (10 times), the gate voltage needs to be reduced by at least 60 mV. In practical transistors, the SS ranges from 70 mV/dec to 100 mV/dec. Increasing the gate-to-channel coupling (e.g., by FinFET structure) will help to lower SS and shrink the off-state leakage current.

Second, the transistor may suffer from the gate-induced-drain-leakage (GIDL) current, which typically occurs when the gate-to-drain potential difference (V_{GD}) or the gate-to-source potential difference (V_{GS}) becomes a negative value. The mechanism for GIDL is the band-to-band tunneling as the carriers may directly tunnel through from the valence band to the conduction band when the energy band diagram is severely bent as the surface of the silicon due to a large negative V_{GD} or V_{GS}. The GIDL current could be observed in the transistor's transfer characteristics (when the V_G becomes negative), as indicated in Figure 2.21(a).

2.3.2 SRAM's Leakage Reduction

Considering the nominal bias condition for the hold operation ($V_{WL} = 0$, $V_{BL} = V_{DD}$), Figure 2.22(a) shows the primary leakage paths of a 6T cell during the hold. Three transistors are susceptible to off-state current, and four transistors are susceptible to

Static Random Access Memory (SRAM)

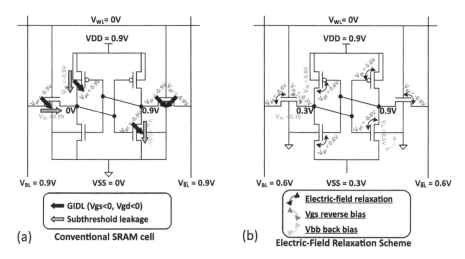

FIGURE 2.22 (a) Primary leakage paths of a 6T cell during hold. (b) Electric field relaxation scheme to minimize the leakage current during hold.

GIDL current. The sum of these currents gives the overall leakage current for one 6T cell, and this leakage current multiplied by the power supply V_{DD} yields the standby power consumption of one 6T cell, which could be in the range of nW. However, considering an MB-level cache, the total standby power consumption could be quite notable in the range of mW.

To minimize the standby power consumption, several approaches are available. First, high-V_{th} transistors could be used with a possible trade-off with slower access speed. Second, an optimized bias condition could be applied during the hold, as shown in Figure 2.22(b). Given the power supply $V_{DD} = 0.9$ V, the lowest level of the cross-coupled inverters could be raised (e.g., $V_{SS} = 0.3$ V) and also the BL and BL voltages could be lowered to the intermediate level (e.g., $V_{BL} = V_{\overline{BL}} = 0.6$ V). The electric field across the drain to the source is relaxed to 1/3–2/3 of the conventional design; thus, the off-state currents through PU1, PG2, and PD2 are suppressed. In addition, the electric field across the gate to the source/drain in such a bias condition is much relaxed to 1/3–2/3 of the conventional design; thus, the GIDL currents through PU1, PG1, PG2, and PD2 are also suppressed.

2.4 VARIABILITY AND RELIABILITY

2.4.1 Transistor Intrinsic Parameter Fluctuations and the Impact on SRAM Stability

One of the great challenges to scale the SRAM technology is the semiconductor manufacturing process variation. The nanoscale transistors may exhibit significant intrinsic parameter fluctuations in V_{th}, I_{off}, and I_{on}. As a result, there will be a mismatch in the 6T cell (between the left and right branches) and cause the asymmetric butterfly curves. The H-SNM or R-SNM will be suppressed as it is defined as the smaller nested square that could fit within the butterfly curves. Figure 2.23 shows

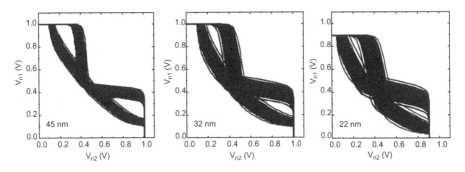

FIGURE 2.23 6T SRAM cell's butterfly curves during read showing variability increases for scaling technology nodes from 45 nm to 28 nm.

examples of the 6T cell's butterfly curves for technology scaling from 45 nm node to 28 nm node. The dispersion of butterfly curves tends to increase with scaling. Less noise margin is available for a more advanced technology node (if using the bulk planar transistor structure). The problem is exacerbated when V_{DD} is further reduced with scaling.

The primary variation sources for nanoscale transistor's intrinsic parameter fluctuations include the random dopant fluctuation (RDF), the line edge roughness (LER), and the metal work function variation (WFV) [6], as shown in Figure 2.24. The number of dopants becomes fewer and fewer when the transistor's channel length (L) shrinks (e.g., less than 100 dopants when $L < 50$ nm). Therefore, the number of dopants and their relative locations may vary from one transistor to another. As the dopant spatially determines the local potential barrier from the source to the channel, the transistor's V_{th} is significantly affected by the RDF effect. In the first order analysis, the sigma of V_{th} distribution (in planar transistors) is inversely proportional to the square root of the gate area (WL) as shown in the following equation:

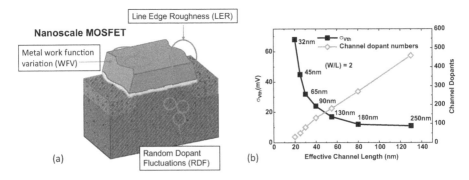

FIGURE 2.24 (a) Primary variation sources for nanoscale transistor's intrinsic parameter fluctuations include the random dopant fluctuation (RDF), the line edge roughness (LER), and the metal work function variation (WFV). (b) The number of dopants and the sigma of threshold voltage as a function of the transistor's channel length.

Static Random Access Memory (SRAM) 41

$$\sigma V_{th(RDF)} = \frac{q}{C_{ox}}\sqrt{\frac{N_d W_{dm}}{3LW}} \qquad (2.3)$$

where q is the basic charge quantity, N_d is the doping concentration of the channel, W_{dm} is the maximum depletion width of the channel to the substrate, and C_{ox} is the gate capacitance (per unit area).

The LER is caused by the imperfect lithography and its following etching process. Due to the molecular nature, the photoresist's edge becomes rough when it is exposed to the light source during the lithography. The rough edge of the photoresist is then further transferred to the underlying pattern (e.g., the gate edge or the fin edge in FinFET), with a typical sigma of roughness 1–2 nm. When the critical dimension of the gate length becomes sub-30 nm (or the fin width is typically sub-10 nm), the LER effect becomes significant. Metal WFV is another concern for the advanced transistors that adopt the high-k/metal gate technology. Metal alloys are usually used to tune the threshold voltage of the transistors. The polycrystalline nature of metal alloys gives rise to the variations, especially when the critical dimension shrinks. Since FinFET uses an undoped or lightly doped channel, the RDF is not a concern. For FinFET, LER and WFV dominate its intrinsic parameter fluctuations. Since RDF, LER, and WFV are regarded as statistically independent randomness sources, their contributions to the sigma of V_{th} could be expressed as the following equation:

$$\sigma V_{th(total)}^2 = \sigma V_{th(RDF)}^2 + \sigma V_{th(LER)}^2 + \sigma V_{th(WFV)}^2 \qquad (2.4)$$

2.4.2 TEMPORAL RELIABILITY ISSUES AND THE IMPACT ON SRAM STABILITY

The intrinsic parameter fluctuation is essentially a static effect that determines the spatial variations from one transistor to another. On the other hand, the transistor also exhibits temporal reliability issues such as random telegraph noise (RTN) [7] and bias temperature instability (BTI) [8]. Here, RTN is a short-term effect, while BTI is a long-term effect.

Figure 2.25 (a) illustrates the RTN effect, as reflected as the V_{th}'s temporal variation (between two levels) when measured using a constant drain current source. When the carriers travel along the channel, there is a certain probability that the carriers are captured by the traps in the oxide, and released from the traps after some time. This will result in fluctuated drain current (or fluctuated threshold voltage if the drain is biased with a current source) over time. When there are many traps, the carrier capture and emission occur frequently, and the drain current may be smoothed out as random noise. However, when there is only one dominant trap along the channel in a nanoscale transistor, the two-level fluctuation becomes distinguishable. Figure 2.25(b) shows an exacerbated V_{th} tail distribution due to RTN when the transistor's dimensions scale down. Such RTN effect will affect the determination of SRAM cell's minimum V_{DD}, and more margin should be allocated for a robust design.

BTI is an effect that V_{th} tends to drift over time under stress which could be accelerated by high temperature. BTI could occur in both NMOS and PMOS transistors, but PMOS suffers more significantly than NMOS in practice. Since PMOS has a

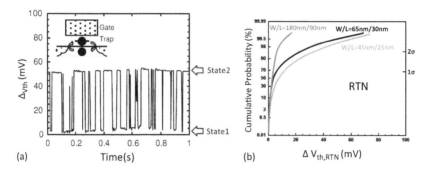

FIGURE 2.25 (a) The random telegraph noise (RTN) mechanism showing threshold voltage (V_{th}) fluctuation over time. (b) V_{th} tail distribution due to RTN when the transistor's dimensions scale down.

negative gate voltage, the effect is thus referred to as NBTI. This is related to the generation of interface traps at the oxide interface when a negative voltage is applied to the PMOS gate for a long time. Figure 2.26(a) shows the NBTI effect for a PMOS transistor. The absolute value of V_{th} tends to increase over time (t) when a large stress gate voltage is applied and such increase could be speeded up at high temperatures. NBTI-induced threshold voltage drift could be empirically fitted with a polynomial function

$$V_{th,\text{NBTI}}(t) = At^n \qquad (2.5)$$

where A and n are coefficients. NBTI will lead to a long-term reliability concern for SRAM. Figure 2.26(b) shows the R-SNM as a function of operational time for different V_{DD}. Therefore, the initial design of SRAM will need to budget enough noise margin for the cell to be functional at a later stage of its lifetime.

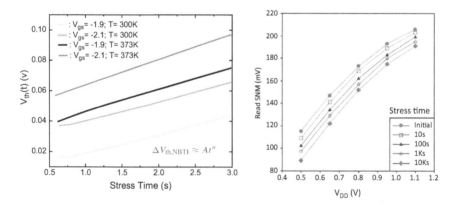

FIGURE 2.26 (a) The negative bias temperature instability (NBTI) effect for a PMOS transistor. (b) Read static noise margin as a function of operational time for different V_{DD}.

2.4.3 SOFT ERROR CAUSED BY RADIATION EFFECTS

SRAM is also sensitive to the radiation effects, which could be classified as single-event-upset (SEU) effect and total-ionizing-dose (TID) effect. SEU occurs when the energetic particles hit the SRAM cell one time and cause the SRAM state to flip in one or multiple cells. TID occurs when energetic particles cumulatively hit the SRAM cell and cause the V_{th} permanent shift in the transistors. Therefore, SEU is a short-term effect, while TID is a long-term effect.

Radiation effects are generally originated from two primary sources. The first source is Alpha particles (He^{2+}), which is typically a nuclear decay of unstable isotopes from package and encapsulation materials. Alpha particles could be shielded, therefore are of less concern. The second source is cosmic ray neutrons and heavy ions, which is challenging to be shielded and can cause much more severe outcome. Cosmic rays are a major concern for aerospace electronics where SRAM cache is used. When the energetic neutrons and heavy ions hit the silicon substrate, they will transfer the energy (higher than the silicon bandgap) to excite the electrons from the valence band to the conduction band. As a result, electron–hole pairs are generated along the tracks of the particles. This causes a severe problem when the strike occurs at the p/n junction of the transistor (e.g., the drain to body junction of the transistor). Because of the built-in potential barrier and built-in electric field, electron–hole pairs will be split, with electrons being collected by higher potential (e.g., drain of an NMOS) and the holes being collected by ground (e.g., the body of an NMOS).

The generated electron–hole pairs could cause a large but short noise current (e.g., in the range of mA and tens of ps) to the storage node of an SRAM cell, which could be modeled as a noise current injection using the dynamic analysis discussed earlier in Figure 2.15(b). The SRAM state thus may be flipped if the storage node "0" is being hit, leading to an SEU. Note that SEU is not a permanent failure, but just a temporary failure; thus, it is termed as a "soft error". The soft error rate tends to increase with technology scaling because the critical charge (Q_{crti}) is reduced due to storage node capacitance reduction when dimensions scale down.

In an SRAM array, multi-bit error may occur even under a single strike [9]. Figure 2.27 shows the mechanism of multi-bit error. Cell_0's storage node "1" is being hit, and it is flipped to "0" as the electrons are injected into its storage node N2. On the other hand, the holes need to sink to the grounded body terminal. The hole current will flow along the p-type substrate to the body contact. Due to the finite resistance of the p-type substrate, the hole current will raise the substrate potential above zero at the V_p point. If V_p is sufficiently high, a parasitic bipolar junction transistor (BJT) between the n-type source contacts of PD transistors of Cell_0 and Cell_1 is turned on. Cell_1's storage node N1 (if stores "1") is thus discharged by such BJT current, causing Cell_1 flip as well. This is the so-called "latch-up" effect, and it may propagate to multiple cells along the way that the substrate potential gradually decays from high to ground. To eliminate such a latch-up effect, silicon-on-insulator (SOI) transistor technology could be used for the radiation-hard SRAM design, because the substrate is isolated by the insulator and there is no path for the hole current to travel along.

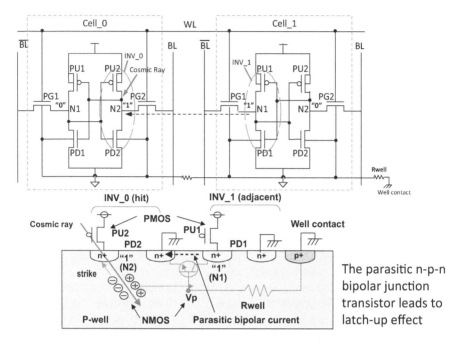

FIGURE 2.27 The mechanism of multi-bit error caused by a single event strike, showing the latch-up effect.

2.5 SRAM LAYOUT AND SCALING TREND

2.5.1 6T Cell Layout

The 6T cell is used as an example to illustrate the general design rule consideration for SRAM layout. Figure 2.28 shows the representative 6T cell layout from the 90 nm node to the 32 nm node using planar transistors. The layout could be interpreted in the following way. The gate is aligned horizontally, and the channel direction is placed vertically. The middle part is the N-well that contains the two pull-up PMOS transistors. Each pull-up PMOS transistor shares the same horizontal gate with the pull-down NMOS transistor. Each pull-down NMOS transistor is connected to the pass-gate NMOS transistor that vertically shares the same drain contact via and occupies the same active region. In this particular example, it is seen that the PD:PG:PU transistor W/L ratio is 2:1:1. Given this layout, the 6T cell area in terms of F^2 could be estimated. Vertically, the unit cell has 2 contacted poly-pitch (CPP, or CGP) distance from the V_{SS} contact to the BL contact. Horizontally, the unit cell has 5 M1 pitch through six contact vias (WL, BL, V_{DD}, V_{DD}, BL, WL). The left and right WL contact vias will be connected on upper-level metal (not shown here). It should be noted that approximately M1 pitch = 2 F, and CPP = 4 F for these technology nodes (90–32 nm) considering the gate length, spacer between gate and drain/source, and drain/source contact size. Therefore, the cell area is given by 8F (vertical) × 20F (horizontal) = 160 F^2. 150–160 F^2 is a typical lower bound of a high-density SRAM 6T cell, and more than 300 F^2 may be required if larger W/L is used for faster access.

Static Random Access Memory (SRAM)

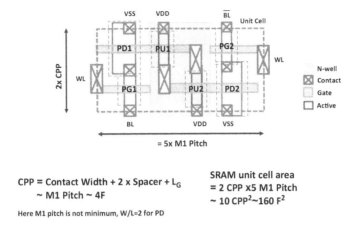

FIGURE 2.28 Representative 6T SRAM cell layout using planar transistors, showing 160 F^2 cell area.

2.5.2 SRAM Scaling Trend

The general SRAM layout has been maintained for many generations in the planar transistor era and the scaling factor is roughly 0.5× in the absolute cell area (in µm^2) as shown by the trend in Figure 2.29. Microscopic top-view images of the fabricated SRAM 6T cell are also shown and the horizontal gate and vertical active region could be seen. SRAM is the most demanding circuit in the logic process that

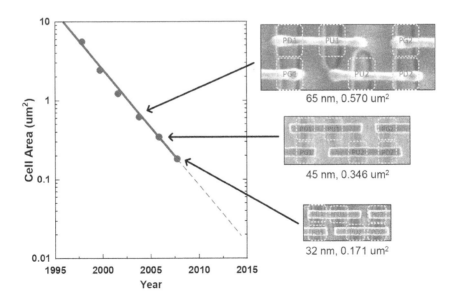

FIGURE 2.29 The SRAM layout scaling of the absolute cell area (in µm^2) and microscopic top-view images of the fabricated SRAM 6T cell in the planar transistor era.

TABLE 2.1
The Recent Trend of SRAM Scaling from 22 nm Node toward 5 nm Node. Here, F Is Assumed to be the Same as the Technology Node Symbol for the Normalization Purpose

Tech node	22 nm	14 nm	10 nm	7 nm	5 nm
Bit cell (µm²)	0.092	0.059	0.031	0.026	0.021
Normalized to F²	190	300	310	530	840

Notes: f is (incorrectly) assumed to be the same as technology node for nomalization purpose.

requires precise lithography patterning. This is because the SRAM layout has been highly optimized by the foundry with minimum distance between the contacts, isolation, and wires that seem to violate the logic design rule. In the 65 nm SRAM cell layout microscopic image, the gate patterns are imperfect with round shape (not straight rectangle), due to the optical proximity effect as the 65 nm is well beyond the theoretical resolution ($\lambda/2$) of photolithography with ArF ultraviolet source (with wavelength $\lambda = 193$ nm). To extend the applicability of 193 nm photolithography for technology nodes 65 nm node and beyond, a variety of techniques are used including the immersion lithography (that increases the numerical aperture of the lens), the optical proximity correction (that tends to compensate the optical imperfection by predictive layout design), and the multiple (double, triple, quadruple) patterning (that split the complex patterns to several simple steps). Going forward to the 7 nm node and beyond, extreme ultraviolet (EUV) lithography that uses a much shorter wavelength of 13.5 nm is a must.

As discussed in Section 1.4, 22 nm is a landmark point where FinFET was introduced and the SRAM cell area shrunk to sub-0.1 µm². Nevertheless, the general 6T layout rule still applies. The new features of FinFET-based SRAM will be elaborated on in Section 2.6. Figure 2.17 shows the scaling trend of a high-density SRAM 6T cell toward a 7 nm node. At the same time, the nominal V_{DD} has been reduced to 0.6 V–0.7 V. Though the scaling trend continues with reduced cell area in its absolute cell area (in µm²), the pace has slowed down in recent years. It is reflected as increasing F² that is normalized to the technology node (if assumed to be F incorrectly[6]). For high-density 6T cell, 160 F² in the 22 nm node becomes 840 F² in the 5 nm node, as shown in Table 2.1.

2.6 FINFET-BASED SRAM

2.6.1 FinFET Technology

As introduced earlier, FinFET is a key enabler for technology scaling beyond the 22 nm node. Figure 2.30 shows the schematic of a planar transistor and a non-planar FinFET. The thin fin is typically etched down from the bulk silicon's surface. FinFET has one or multiple thin fins that conduct the current (that still flows at the interface of silicon and gate oxide). The FinFET typically has a tri-gate structure where the gate

Static Random Access Memory (SRAM) 47

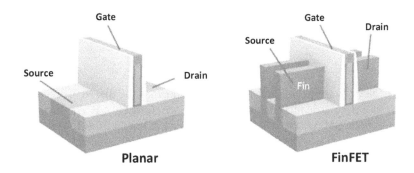

FIGURE 2.30 The schematic of a planar transistor and a non-planar FinFET.

is covering both sides of the fin and the top of the fin. The enhanced gate-to-channel coupling helps mitigate the short channel effect. The FinFET concept was first experimentally proved by Prof. Chenming Hu's group at UC Berkeley in 1998 [10].

After 14 years of efforts by academic and industrial research and development, Intel was the first company to announce its commercialization at the 22 nm node in 2012. This FinFET is based on a bulk silicon with high-k/metal-gate and gate-last fabrication progress. Figure 2.31 shows the microscopic images of Intel's 22 nm FinFET technology [11]. Two slices are cut along the gate direction (A-A'), and along the fin direction (B-B'), and their cross-section views are shown. From the A-A' cross section, it is seen that in this 22 nm process, the fin width is about 8 nm and the fin height is about 34 nm. As there are three sides of the fin that are

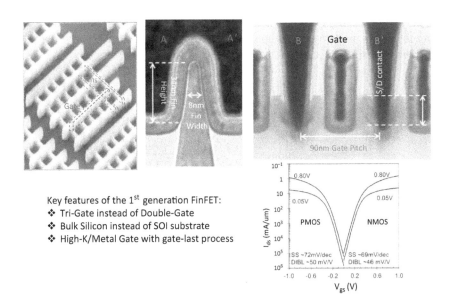

FIGURE 2.31 Microscopic images of Intel's 22 nm FinFET technology, and the I_D-V_G characteristics of both NMOS and PMOS.

interfaced with the gate, the effective electrical width that conducts the current now is defined as

$$W_{\text{eff}} = 2 \times Fin_{\_height} + Fin_{\_width} \quad (2.6)$$

In this case, W_{eff} is 76 nm for a single fin. As shown in the B-B' cross section, CGP is 90 nm along the channel direction, which is approximately 4 F. The ID-V_G characteristics of both NMOS and PMOS are also shown in Figure 2.31. Excellent electrical parameters such as ~70 mV/dec SS and ~50 mV/V DIBL[7] factors are obtained. It is also seen that the NMOS and the PMOS current drivability is similar, which suggests that the mobility advantage of electrons over holes in planar transistors diminishes in the FinFET era.

For circuit designers, the major change brought by the FinFET technology is the quantized electrical width on the layout, as shown in Figure 2.32. As FinFET can only offer a discrete number of fins, its electrical width can only be $N \times W_{\text{eff}}$ (N is an integer, 1, 2, 3...). Therefore, less flexibility is allowed for the transistor sizing. Another challenge of FinFET is that the V_{th} modulation is less straightforward as varying doping concentration is no longer effective with ultra-thin fin (typically undoped). Instead, metal work function engineering is often used for the V_{th} modulation. On the other hand, FinFET provides an opportunity to improve the integration density as now the fin pitch no longer determines the current drivability. Physically occupying one fin pitch (i.e., 60 nm in Intel's 22 nm node) in the lateral footprint, W_{eff} is 76 nm per fin. Hence, it represents a boost of 1.27× current density in the lateral footprint.

The scaling trend of FinFET dimensional parameters from the 22 nm node to the 7 nm node is summarized in Table 2.2. It shows that the fin becomes thinner, taller, and denser, reflected as reduced fin width, increased fin height, and smaller fin pitch. The current drivability boosting factor becomes 3.67× at the 7 nm node. This allows using fewer fins to deliver the same amount of current, thus further improving the integration density as discussed in Section 1.4.

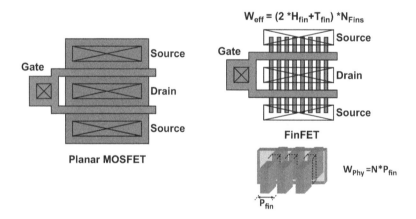

FIGURE 2.32 Layout differences between the planar transistor and the FinFET.

TABLE 2.2
The Scaling Trend of FinFET Dimensional Parameters from 22 nm to 7 nm

	22 nm	14 nm	10 nm	7 nm
Fin_height	34	37	42	52
Fin_width	8	8	6	6
Fin_pitch	60	48	36	30

2.6.2 SRAM Scaling in FinFET Era

FinFET offers many advantages for SRAM design to advanced technology nodes. For example, the improved SS allows using lower V_{th} at given I_{off}, yielding higher current (through PG transistor) for faster read and write. The reduced DIBL leads to a larger output resistance of the transistor and thus, shaper transition in the butterfly curve for larger R-SNM. Figure 2.33 shows a comparison of the butterfly curves during a read for planar transistor-based SRAM and FinFET-based SRAM. The undoped channel in FinFET also minimizes RDF-induced variability.

Intel's 22 nm SRAM family is shown in Figure 2.34 with different cell options [12], by varying the fin numbers in the PD, PG, and PU transistors. The high-density uses a compact 1(PD):1(PG):1(PU) ratio for a low-power (LP) design. The standard cell uses a moderate 2 (PD):1(PG):1(PU) for standard performance (SP) design. The high-speed cell uses a relaxed 3(PD):2(PG):1(PU) design for high performance (HP) design. Therefore, the different cells could implement SRAM arrays with different characteristics, as shown in the Shmoo plot by varying the V_{DD} and clock frequency. Though HP design could achieve 4.6 GHz frequency at $V_{DD} = 1$ V, it has tradeoffs with high dynamic power and standby power.

Figure 2.35 summarizes the key specifications of the high-density SRAM cells from major manufacturers in the leading-edge nodes. As discussed earlier, the

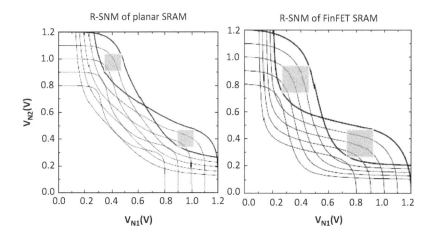

FIGURE 2.33 Butterfly curves during the read for planar transistor-based SRAM and FinFET-based SRAM.

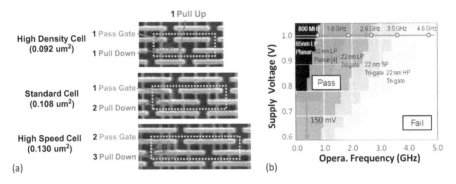

FIGURE 2.34 (a) Intel's 22 nm FinFET SRAM family with different cell options. (b) Shmoo plot of the SRAM performance improvement over generations and between cell options.

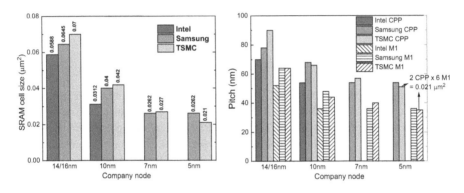

FIGURE 2.35 Key specifications of the high-density SRAM cells from major manufacturers in the leading-edge nodes.

technology node is at discretion of the company. Hence, even at the same node, the SRAM cell size varies among different manufacturers. It is reasonable to say that Intel's 10 nm process is close to TSMC's 7 nm process in terms of integration density. Looking forward to the future SRAM design, the stacked nanosheet transistor technology could potentially offer a more compact layout and design flexibilities by varying the number of stacked layers for PD, PG, and PU.

NOTES

1. If not explicitly specified, SRAM discussed in this book refers to such 6T cells. If not explicitly specified, transistor discussed in this book refers to the metal-oxide-semiconductor-field-effect transistor (MOSFET).
2. In most planar transistors, electrons have a significantly higher (e.g., twice) mobility than holes. However, the hole mobility is approaching the electron mobility in the FinFET era.
3. Here, N2 is assumed to switch from 0 to 1, while in the earlier example in Figure 2.5, N1 is assumed to switch from 0 to 1.

4 Here, the gate oxide is assumed to be SiO_2, and Si's permittivity is about 3 times that of SiO_2.
5 $\ln(10) = 2.3$ is the coefficient that transfers from the logarithm base 10 to base e. kT/q is the thermal voltage determined by Boltzmann statistics, and it is 26 mV at room temperature.
6 Here, F is simply assumed to be the same as the technology node (22 nm, 14 nm, 10 nm, 7 nm, and 5 nm) for normalization purpose. As discussed in Chapter 1 (let's make it more specific, change to Section 1.4), this is an incorrect representation as the node name is a symbol that does not correlate with the physical dimensions.
7 DIBL is drain-induced-barrier-lowering effect that shows the impact of drain voltage on threshold voltage. The DIBL factor is defined as the V_T shift over the drain voltage change between a linear region (e.g., with $V_{DS} = 50$ mV) and a saturation region (e.g., with $V_{DS} = V_{DD}$).

REFERENCES

[1] J. Wang, S. Nalam, B.H. Calhoun, "Analyzing static and dynamic write margin for nanometer SRAMs," *IEEE International Symposium on Low Power Electronics and Design (ISLPED)*, 2008, pp. 129–134. doi: 10.1145/1393921.1393954
[2] E. Grossar, M. Stucchi, K. Maex, W. Dehaene, "Read stability and write-ability analysis of SRAM cells for nanometer technologies," *IEEE Journal of Solid-State Circuits*, vol. 41, no. 11, pp. 2577–2588, November 2006. doi: 10.1109/JSSC.2006.883344
[3] W. Dong, P. Li, G. Huang, "SRAM dynamic stability: theory, variability and analysis," *IEEE/ACM International Conference on Computer-Aided Design (ICCAD)*, 2008, pp. 378–385. doi: 10.1109/ICCAD.2008.4681601
[4] M. Khellah, Y. Ye, N.S. Kim, D. Somasekhar, G. Pandya, A. Farhang, K. Zhang, C. Webb, V. De, "Wordline & bitline pulsing schemes for improving SRAM cell stability in low-Vcc 65nm CMOS designs," *IEEE Symposium on VLSI Circuits*, 2006, pp. 9–10. doi: 10.1109/VLSIC.2006.1705286
[5] L. Chang, R.K. Montoye, Y. Nakamura, K.A. Batson, R.J. Eickemeyer, R.H. Dennard, W. Haensch, D. Jamsek, "An 8T-SRAM for variability tolerance and low-voltage operation in high-performance caches," *IEEE Journal of Solid-State Circuits*, vol. 43, no. 4, pp. 956–963, April 2008. doi: 10.1109/JSSC.2007.917509
[6] A. Asenov, "Simulation of statistical variability in nano MOSFETs," *IEEE Symposium on VLSI Technology*, 2007, pp. 86–87. doi: 10.1109/VLSIT.2007.4339737
[7] N. Tega, H. Miki, F. Pagette, D.J. Frank, A. Ray, M.J. Rooks, W. Haensch, K. Torii, "Increasing threshold voltage variation due to random telegraph noise in FETs as gate lengths scale to 20 nm," *IEEE Symposium on VLSI Technology*, 2009, pp. 50–51.
[8] S. Bhardwaj, W. Wang, R. Vattikonda, Y. Cao, S. Vrudhula, "Predictive modeling of the NBTI effect for reliable design," *IEEE Custom Integrated Circuits Conference (CICC)*, 2006, pp. 189–192. doi: 10.1109/CICC.2006.320885
[9] K. Osada, K. Yamaguchi, Y. Saitoh, T. Kawahara, "Cosmic-ray multi-error immunity for SRAM, based on analysis of the parasitic bipolar effect," *IEEE Symposium on VLSI Circuits*, 2003, pp. 255–258. doi: 10.1109/VLSIC.2003.1221220
[10] D. Hisamoto, W.-C. Lee, J. Kedzierski, E. Anderson, H. Takeuchi, K. Asano, T.-J. King, J. Bokor, C. Hu, "A folded-channel MOSFET for deep-sub-tenth micron era," *IEEE International Electron Devices Meeting (IEDM)*, 1998, pp. 1032–1034. doi: 10.1109/IEDM.1998.746531

[11] C. Auth, C. Allen, A. Blattner, D. Bergstrom, M. Brazier, M. Bost, M. Buehler, et al., "A 22nm high performance and low-power CMOS technology featuring fully-depleted tri-gate transistors, self-aligned contacts and high density MIM capacitors," *IEEE Symposium on VLSI Technology*, 2012, pp. 131–132. doi: 10.1109/VLSIT.2012.6242496

[12] C.-H. Jan, U. Bhattacharya, R. Brain, S.-J. Choi, G. Curello, G. Gupta, W. Hafez, et al. "A 22nm SoC platform technology featuring 3-D tri-gate and high-k/metal gate, optimized for ultra low power, high performance and high density SoC applications," *IEEE International Electron Devices Meeting (IEDM)*, 2012, pp. 3.1.1–3.1.4. doi: 10.1109/IEDM.2012.6478969

3 Dynamic Random Access Memory (DRAM)

3.1 DRAM OVERVIEW

3.1.1 DRAM Sub-system Hierarchy

Standalone DRAM is commonly used as the main memory for a computer system. Figure 3.1 shows the DRAM sub-system and its hierarchy for a representative workstation or server. Multiple channels communicate with the processor through the 64-bit wide data bus for each channel. The dual in-line memory module (DIMM) is a circuit board that integrates multiple (e.g., 8) DRAM chips with 8-bit input/output (I/O) width from each chip. One DRAM chip consists of many banks; 8 bits are multiplexed between multiple banks. Within one bank, there are many sub-arrays, namely mats. One mat consists of the memory cell array and peripheral circuits such as row/column decoders, column MUX, sense amplifiers, etc. Take a 4 GB DRAM chip as an example; there might be 16 banks (each bank is 256 MB), and one bank may contain 16 × 16 mats (each mat is 1 MB), and one mat size could be 1024 × 1024 B. Note here one Byte is for 8 bits as a group without further column decoding.

3.1.2 DRAM I/O Interface

DRAM chips have specially designed input/output (I/O) interfaces following a certain protocol. One of the most widely used interfaces is the double-data-rate (DDR) series, which is a synchronous mechanism to boost the data transfer rate given a limited DRAM internal clock frequency. For example, the DRAM internal clock cycle time is 5 ns (thus the frequency is 200 MHz) due to the required time for correct sensing. The interface uses double pumping (transferring data on both the rising and falling edges of the clock signal), and thus is referred to as the double data rate. DDR effectively doubles the I/O bandwidth at the same internal clock frequency, achieving 400 M bit per second per I/O pin (i.e., 400 Mbps/pin), and thus 3.2 GB/s for a 64-bit-wide DIMM. Since its invention, the DDR protocol has extended to several generations. For example, DDR2, DDR3, DDR4, and DDR5 simply means the interface clock frequency will be running at a 2×, 4×, 8×, and 16× higher frequency than the DRAM internal clock frequency, resulting in substantial boosts of the I/O bandwidth. Take an example of the DDR3; if the DRAM internal clock frequency is

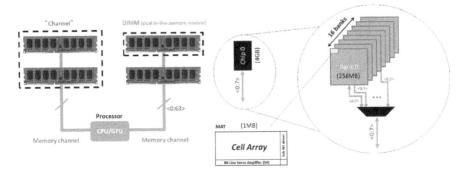

FIGURE 3.1 The DRAM sub-system and its hierarchy including the channel, DIMM, chip, bank, and mat.

200 MHz, and the interface clock frequency is 800 MHz, it results in 1600 Mbps/pin, and thus 12.8 GB/s for a 64-bit-wide DIMM.

What essentially the DDR interface does is to serialize/de-serialize between the DRAM internal core and the I/O interface to the data bus, as shown in Figure 3.2. Take DDR3 as an example again; 8 bits are prefetched from multiple columns in one internal clock cycle and stored in prefetch buffer, and they are burst out with 8× higher rate at one pin of the I/O in one interface clock cycle.

The DDR protocol has a family of variants including LPDDR and GDDR. LPDDR is for low-power mobile platforms, and it is optimized for low leakage with lower supply voltage and long refresh interval. GDDR is for high-performance graphics platforms, and it is optimized for fast internal clock frequency with a small mat array size. Figure 3.3 shows the trends of I/O bandwidth for different interface protocols, including DDR, LPDDR, and GDDR.

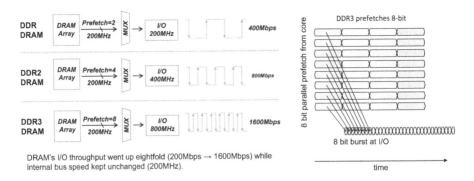

FIGURE 3.2 Illustration of DDR protocol that prefetches bits in parallel and burst with higher I/O clock frequency to increase the I/O bandwidth.

Dynamic Random Access Memory (DRAM)

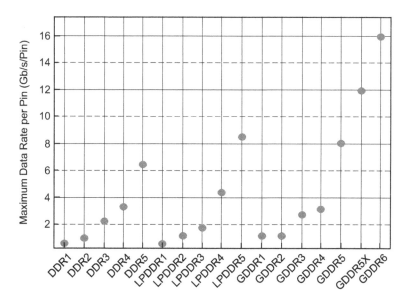

FIGURE 3.3 The trends of I/O bandwidth for different DRAM interface protocols, including DDR, LPDDR, and GDDR.

3.2 1T1C DRAM CELL OPERATION

3.2.1 Principle of 1T1C Cell

DRAM has a bit cell of 1-transistor-1-capacitor (1T1C). Figure 3.4(a) shows the 3D schematic of today's representative 1T1C DRAM bit cell and its periphery. Two kinds of transistor technologies exist in the DRAM process: the cell access transistor that controls the access to the storage node capacitor and the peripheral transistor that is used to build the peripheral circuits such as decoders and sense amplifiers. Figure 3.4(b)

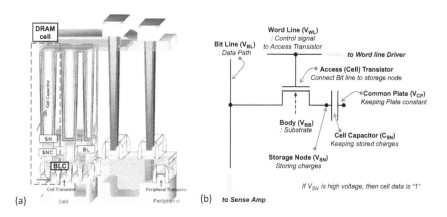

FIGURE 3.4 (a) 3D schematic of the representative 1T1C DRAM bit cell and its periphery. (b) Circuit schematic of the 1T1C bit cell.

shows the circuit schematic of the 1T1C bit cell. The cell access transistor's gate is controlled by the word line (WL). One contact of the cell transistor's source/drain is connected to the bit line (BL), and the other contact is connected to one electrode of the storage node capacitor. The storage node (SN) capacitor is physically stacked on top of the cell transistor with a high aspect ratio. The other electrode of the SN capacitor is attached to a common plate (CP). Typically, two bit cells share one contact via to the BL.

DRAM relies on the charges that are stored on the SN capacitor for data storage. If there are sufficient charges on the SN capacitor, and the SN voltage is high (i.e., V_{DD}), then it stores "1"; if there is no charge on the SN capacitor, and the SN voltage is low (i.e., ground), then it stores "0". To write the data into the 1T1C bit cell is straightforward. "1" is written by biasing BL to V_{DD} while turning on the WL with V_{PP}, typically a higher voltage than V_{DD} to facilitate the current passing through the cell transistor to charge the SN capacitor up to V_{DD}. "0" is written by biasing BL to ground while turning on the WL with V_{PP}, typically a higher voltage than V_{DD} to facilitate the current passing through the cell transistor to discharge the SN capacitor down to the ground. During the hold operation, WL is turned off; thus, the charging state could maintain on the SN capacitor for some period of time. Due to the existence of leakage paths, the charging state will decay over time. Therefore, a periodic refresh is required for DRAM to maintain its memory state (which is different from SRAM). If the power supply is removed, the memory state is lost in DRAM. Therefore, DRAM is a volatile memory (same as SRAM).

3.2.2 Charge Sharing and Sensing

Reading the memory state from the 1T1C bit cell involves a charge-sharing process between the SN capacitance (C_{SN}) and the BL capacitance (C_{BL}). It is assumed that the SN stores "1" and its voltage is V_{DD}. First, the BL is pre-charged to $V_{DD}/2$. Then, WL is turned on with V_{PP}. Since the SN's potential is higher than that of the BL in this case, current (or the "positive" charges[1]) will flow from SN to BL through the cell access transistor. As a result, the SN voltage (V_{SN}) decays while the BL voltage (V_{BL}) is raised. Eventually, V_{SN} equals V_{BL} after the charge sharing is complete. The V_{BL} is thus increased from the original $V_{DD}/2$ by ΔV, which becomes the sense margin. Similarly, if it is assumed that the SN stores "0", the V_{BL} will be decreased from the original $V_{DD}/2$ by ΔV. Positive ΔV indicates the cell stores "1" and negative ΔV indicates the cell stores "0", which will be detected by the sense amplifier.

The sense margin ΔV could be analytically derived as follows using the charge conservation rule. Initially before WL is turned on, SN and BL are separated as the cell access transistor is switched off. The sum of the charges on C_{SN} and C_{BL} is given by

$$C_{sum} = C_{SN} \times V_{DD} + C_{BL} \times V_{DD}/2 \qquad (3.1)$$

After WL is turned on and the cell transistor is switched on, eventually when the charge transfer is complete, SN and BL reach the same potential V_{BL_t}', where BL_t is the true signal line for BL that is connected to the selected bit cell. The circuit has total charges as

Dynamic Random Access Memory (DRAM)

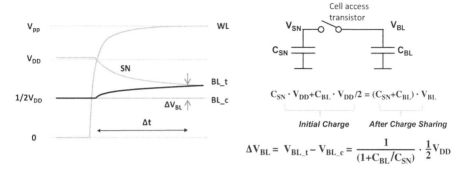

FIGURE 3.5 The waveform for the charge-sharing process in DRAM cell if it stores "1". The simplified two-capacitor model to derive the voltage sense margin ΔV.

$$C_{sum} = (C_{SN} + C_{BL}) \times V_{BL\_t'} \qquad (3.2)$$

Since charges are preserved, C_{sum} should be the same in Equation 3.1 and 3.2; thus, $V_{BL\_t'}$ could be solved and the ΔV is defined as $V_{BL\_t'} - V_{BL\_c}$, where BL_c is the complementary BL that is used as the reference and is biased at $V_{DD}/2$. Therefore, ΔV is given by

$$\Delta V = \frac{1}{1 + C_{BL}/C_{SN}} \times \frac{V_{DD}}{2} \qquad (3.3)$$

As shown in Equation (3.3), the ratio of C_{BL} over C_{SN} is important to determine ΔV. To achieve a larger sense margin that overcomes the process variations and temporal noise, C_{SN} should be maintained sufficiently large while C_{BL} should be minimized.

The DRAM read speed is mainly determined by the cell access transistor's current drivability to complete the charge-sharing process, which could be approximated as

$$\Delta t = C_{BL} \times \Delta V / I_{access} \qquad (3.4)$$

For a typical BL parasitic capacitance C_{BL} = 100 fF, to achieve a sense margin ΔV = 200 mV within Δt = 2 ns, the cell access transistor's drive current I_{access} = 10 µA. It is noted that DRAM has a destructive read nature, as the V_{SN} is changed from its original value during the charge sharing. Hence, a write-back process is required to avoid the read disturb. The write-back is naturally assisted by the sense amplifier, as shown in Figure 3.6(a). A typical latch-based sense amplifier is used for the DRAM sensing, and BL_t (the true line) and the other reference node (the complementary line) are pre-charged to $V_{DD}/2$. When the charge sharing creates sufficient sense margin, the sense amplifier is enabled by the SAEN signal thus is connecting with V_{DD} and the ground. Since the BL is now seeing $V_{DD}/2 + \Delta V$ while the reference is kept at $V_{DD}/2$, the latch starts flipping and the BL is forced to V_{DD} if ΔV is positive. As BL ramps up toward V_{DD}, the cell access transistor is still turned on by WL; thus, the SN capacitor tends to restore "1" by charging back from BL. Or in the other case, if ΔV is negative, BL is forced to ground by the latch. As the cell access transistor is still turned on by WL, the SN capacitor tends to restore "0" by discharging through BL. The waveform for the write-back process is shown in Figure 3.6(b). After the data are restored, WL could be turned off and BL could be pre-charged to $V_{DD}/2$ again.

Figure 3.7 shows the complete set of the peripheral circuits on the column side beside the latch-based sense amplifier. Similar to SRAM peripheral circuits, the 3-transistor equalizer circuit serves for the pre-charge (but to $V_{DD}/2$ for DRAM). The connect circuit helps with passing data in and out to an external buffer outside of the mat array. Figure 3.8 shows the core-timing diagram for a read-modify-write operation and parameters that are typically shown in DRAM's datasheet. Notable parameters include the row cycle time (t_{RC}) for the entire period when the WL is kept on, the row to column delay (t_{RCD}) before the sensing is complete, the write recovery time (t_{WR}) before the cell is being fully written, and the row pre-charge time (t_{RP}) for the BL being pre-charged to $V_{DD}/2$. The CSL pulse that controls the connect circuit determines the internal DRAM clock frequency. For example, 2.5 ns for CSL pulse itself and 2.5 ns for the interval between the read CSL and the write CSL gives a clock cycle of 5 ns, which is translated to a 200 MHz internal DRAM clock frequency (as discussed in the DDR protocol).

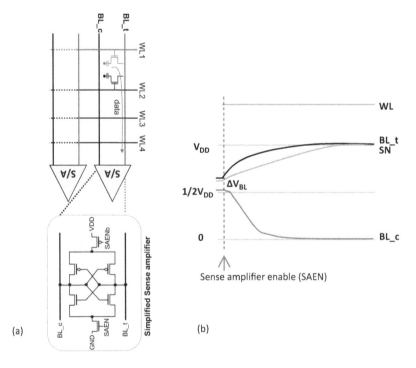

FIGURE 3.6 (a) The write-back process of DRAM with assistance from the latch-based sense amplifier. (b) The waveform for the write-back progress in the DRAM cell.

Dynamic Random Access Memory (DRAM)

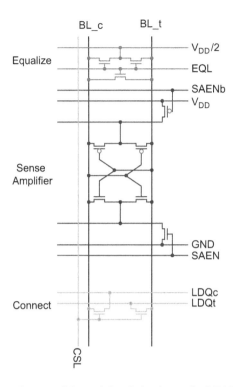

FIGURE 3.7 The complete set of the peripheral circuits on the DRAM column side.

3.2.3 DRAM Leakage and Refresh

A unique feature of DRAM is the dynamic refresh as the charging state in the SN capacitor tends to decay over time due to the leakage paths. Figure 3.9 summarizes the primary leakage current mechanisms of the 1T1C bit cell if it stores "1": (1) The reversed p/n junction between the drain and the body of the cell access transistor introduces junction leakage current (by the drift of minority carriers). (2) The GIDL current due to the band-to-band tunneling that occurs between the drain and the channel under a strong negative gate-to-drain voltage (similarly as discussed in Section 2.3 as one of the SRAM leakage mechanisms). (3) The cell access transistor's off-current due to the subthreshold leakage current (by diffusion of majority carriers over the potential barrier from source to channel) under non-zero drain-to-source voltage. (4) The direct tunneling current through the thin or non-perfect dielectric of the SN capacitor. All of these leakage current paths (I_{leak}) contribute to discharging the SN capacitor, and V_{SN} will decay over time. Therefore, DRAM needs a periodical refresh with a refresh cycle time (t_{REF}) before any read error occurs. The sense margin ΔV of reading "1" is thus reduced by the loss of charge $t_{REF} \times I_{leak}$, as modified from Equation (3.3),

$$\Delta V = \frac{1}{1 + C_{BL}/C_{SN}} \times \left(\frac{V_{DD}}{2} - \frac{I_{leak} t_{REF}}{C_{SN}} \right) \quad (3.5)$$

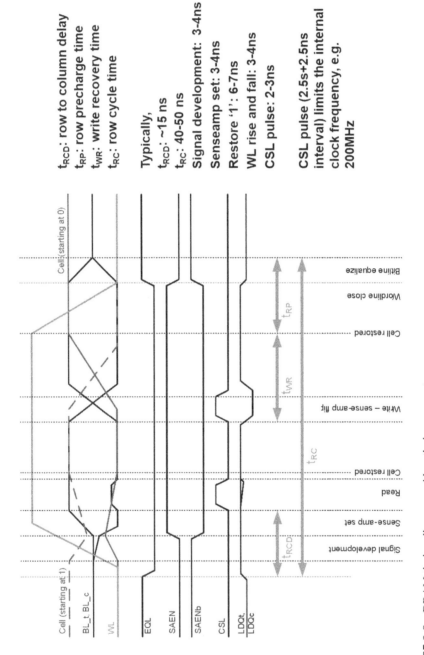

FIGURE 3.8 DRAM timing diagram and key timing parameters.

Dynamic Random Access Memory (DRAM)

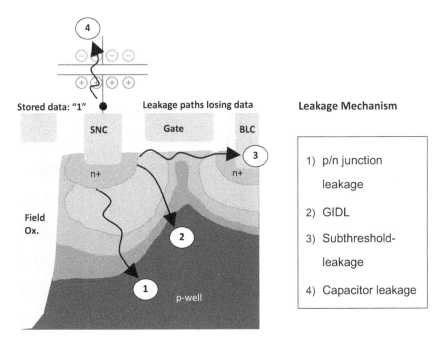

FIGURE 3.9 The primary leakage current mechanisms of the DRAM cell. The iso-electrical potential contours are shown.

Given the required minimum ΔV, then the desired retention time t_{RET} could be back-calculated as,

$$t_{REF} = \frac{C_{SN}}{I_{leak}} \times \left(\frac{V_{DD}}{2} - \left(1 + \frac{C_{BL}}{C_{SN}}\right) \Delta V \right) \quad (3.6)$$

In practice, the refresh cycle time is determined by the tail bits of the retention time distribution measured from the entire DRAM array, as there are significant variations in leakage currents across the bit cells. Figure 3.10 shows an example of the retention time distribution where the majority cells (>99.9%) of the cells have retention time higher than 1 s. However, even 10^{-6} probability of a GB chip will result in hundreds of cells that may have retention time less than 100 ms. Therefore, a widely used industrial standard by the JEDEC [1] is to refresh the DRAM bit cell by every 64 ms.

In practice, DRAM retention characteristics depend on the data pattern and operation modes, as shown in Figure 3.11. To minimize the off-state current through the cell access transistor, the substrate bias (V_{BB}) could be negatively biased. Storing "1" suffers more retention degradation than storing "0", because the SN voltage at V_{DD} introduces more p/n junction reverse bias (with respect to V_{BB}) thus more leakage. Storing "1" could be further classified for static retention mode and dynamic retention mode. The static retention means that all the WLs in the DRAM array are off for the hold, and all the BLs are pre-charged to $V_{DD}/2$. In this scenario, the off-current through cell access transistor could be much suppressed if WL is biased at negative V_{BBW}, then

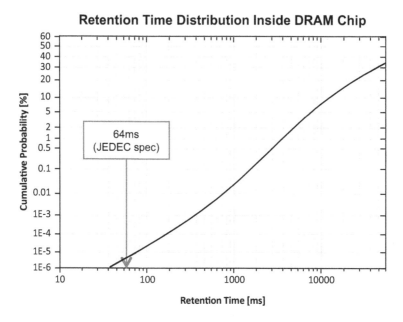

FIGURE 3.10 Retention time distribution measured from a DRAM chip showing the tail bits falling into the 64 ms refresh cycle time.

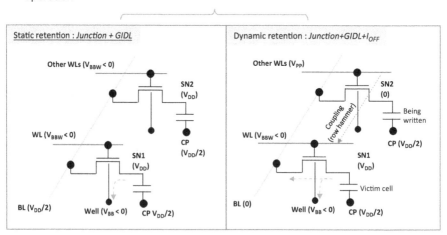

FIGURE 3.11 Static and dynamic retention modes of DRAM cell that store "1". Row hammer effect is the coupling effect between the victim row and the adjacent row that is being activated.

Dynamic Random Access Memory (DRAM)

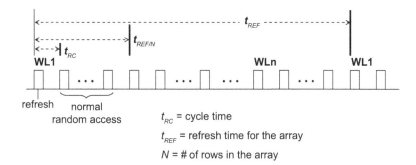

FIGURE 3.12 The DRAM refresh schedule and timing.

the primary leakage paths are the GIDL current as V_{GD} is $|V_{BBW}|$ as well as the p/n junction reversed current with $|V_{BB}|$ negative bias across it. The dynamic retention means that the victim row is in the hold mode but there is an adjacent row being activated. For example, if the adjacent row is being written to "0", then $V_{BL} = 0$, which will exacerbate the off-state current for the victim row with $V_{DS} = V_{DD}$; if the adjacent row is being read repeatedly with WL toggling between V_{PP} and ground, the capacitive coupling between WLs may induce increased leakage of this victim row. This phenomenon is also called the row hammer effect [2].

The refresh in DRAM is in fact performed by the read operation (rather than the write operation) because the write-back progress is built into the read. The memory states could be restored after the read. Hence, the periodic refresh requires reading a specific row of DRAM cells after a time interval t_{REF}. Figure 3.12 shows the timing diagram of the DRAM refresh. It is assumed in this example that the DRAM bank has 16k # of rows and the refresh cycle t_{REF} is 64 ms. A common procedure is to distribute the refresh events uniformly across the refresh cycle. This means the time interval between refreshing one row to the adjacent row $t_{INTERVAL} = t_{REF}/\#$ of rows = 4 μs. If this DRAM row cycle time (t_{RC}) is 40 ns, the time that is required to read a specific row is the same as 40 ns. The refresh overhead is thus $t_{RC}/t_{INTERVAL} = 1\%$, indicating that 1% of the operational time is spent on the refresh. The rest 99% of the operational time could be normal read or write random access to other rows as defined by the input address. But at the predefined time interval, the controller must switch back to the specific row for the read operation according to the refresh scheduling. In a modern DRAM system, the refresh overhead could be more than 20% (not only in operational time, but also in energy consumption).

3.3 DRAM TECHNOLOGY

3.3.1 Trench Capacitor and Stacked Capacitor

There are two major fabrication approaches for the DRAM cell capacitor: the trench capacitor and the stacked capacitor, as shown in Figure 3.13. As discussed earlier, maintaining sufficient SN capacitance is critical to the sense margin. As a rough guideline, C_{SN} = 10 fF–40 fF is required given representative BL parasitic

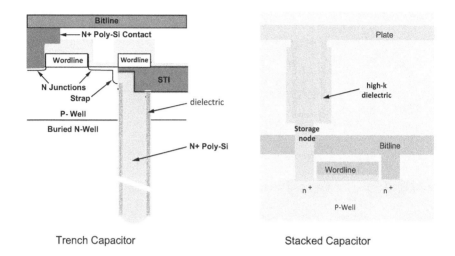

FIGURE 3.13 Schematic of the trench capacitor and the stacked capacitor for DRAM.

capacitance. Nevertheless, the SiO_2-based dielectric gives the capacitance density about 1 fF/µm². To deliver tens of µm² surface area of SN capacitance, the only method is to develop a 3D vertical cylinder-like capacitor as the lateral dimension scaling is toward the nanoscale.

Historically before scaling till 70 nm node, the deep trench capacitor was the mainstream technology for DRAM. The trench capacitor is first processed underneath one of the source/drain contacts of the cell access transistor by digging a deep trench into the silicon substrate. The dielectric is deposited covering the sidewall of the trench, and the inner electrode is refilled with heavily doped poly-silicon. Then the cell access transistor is fabricated on the substrate surface followed by the regular front-end-of-line (FEOL) processing. There are a few challenges for scaling trench capacitor. First, it needs an 8 F^2 cell area prohibiting the 6 F^2 dense cell design rule. Second, there are fabrication issues such as continued increasing aspect ratio of the deep trench, difficulty in the conformal deposition of high-k dielectric in the deep trench, and increased resistance of the trench refill.

The standalone DRAM industry switched to the stacked capacitor in the mid-2000s to enable the scalability toward the 1α nm node as of 2020.[2] The access transistor is fabricated first, and then an SN capacitor is stacked on top of the contact via of the source/drain to allow WL and BL to pass through. Similarly, a 3D vertical cylinder-like capacitor is required. The top common plate is connected by metal wire. The primary motivation for the stacked capacitor is to allow the cell area to be reduced to 6 F^2, and make it easier to stack an even higher-aspect-ratio cylinder with high-k dielectric (enabled by advanced etch and atomic layer deposition). On the downside, the stacked capacitor makes it is difficult to route the metal lines at the back-end-of-line (BEOL) thus leading to poor compatibility with the logic process. Therefore, for embedded DRAM, a deep trench capacitor is still being used (e.g., by IBM).

Dynamic Random Access Memory (DRAM)

3.3.2 DRAM Array Architecture

The DRAM's layout design rule is highly related to its array architecture and capacitor technology. There are two commonly used array architectures depending on how BLs are connected to the sense amplifiers, as shown in Figure 3.14. The first one is called the "folded bit line" architecture. A pair of two BLs from the same array is connected to one sense amplifier. The true (t) line is used for sensing the data, and the complementary (c) line is used as the reference of the sense amplifier which should be kept at $V_{DD}/2$ during sensing. Therefore, the c-line should not be activated by the same WL that controls the t-line. As a result, there should be an empty space where the WL and the c-line intersects. In other words, the DRAM cells are staggered in the array, yielding an 8 F^2 cell area. The second one is called the "open bit line" architecture. Two BLs from two neighboring arrays as t-line and c-line are connected to one sense amplifier that is sandwiched in between. When one array is activated, the other array should be kept on hold. Therefore, the c-line will be kept at $V_{DD}/2$. Since there is no interference from the c-line on the activated array, WL could control all the t-lines in the array without empty space. Two adjacent t-lines could be sensed from two opposite ends of the array and such configuration helps to increase the pitch for sense amplifier layout. The open bit line architecture is now the mainstream option to allow the scalability of cell area to 6 F^2 and potentially 4 F^2 in the future. One drawback of the open bit line architecture is the increased sensitivity toward noise, because the common mode noise from the same array is rejected by the differential input of the sense amplifier in the folded bit line architecture.

Folded bit line architecture

+ Common mode rejection, bit line twist (reduced sensitivity to capacitive coupling), more space for sense amplifier available
- Minimum cell area 8 F^2 (F is minimum feature size, i.e. line width or spacing)

Open bit line architecture

+ Minimum cell area 4 F^2 (6F^2 to fit contact to planar access transistor)
- Larger noise sensitivity

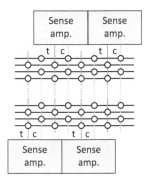
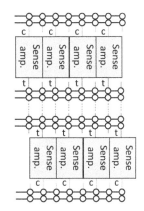

FIGURE 3.14 Schematic of folded bit line and open bit line array architectures of DRAM.

3.3.3 DRAM LAYOUT

The design rule of DRAM layout in the era of the trench capacitor is shown in Figure 3.15. Typically, two bit cells share one BL contact via. Figure 3.15 shows the layout of a folded bit line architecture with the trench capacitor. Since the BL pitch is 2 F and the WL pitch is 4 F (due to the staggered cell location), 8 F^2 cell area is achieved.

One constraint of the stacked capacitor layout design is that the WL or BL could not intersect with the SN capacitor region, as it is standing up vertically. Therefore, a titled cell transistor's channel direction with respect to BL (or WL) is required to leave a space for the vertical 3D stacked capacitor. Figure 3.16 shows today's mainstream layout options of designing the open bit line architecture with the stacked capacitor. In Figure 3.16(a), the BL pitch is 3 F and the WL pitch is 2 F; thus, 6 F^2 cell area is achieved. In Figure 3.16(b), the BL pitch is 2 F and the WL pitch is 3 F (on average, as there is isolating WL as dummy line); thus 6 F^2 cell area is achieved. These designs offer the pros such as straight lines for easier lithography patterning, but the cons are isolated and small active regions (only for two bit cells that share the same BL contact). A variant of the design is shown in Figure 3.16(c), where the BLs and active region are twisted but both are continuous. Still, the BL pitch is 2 F and the WL pitch is 3 F (on average); thus, 6 F^2 cell area is achieved.

3.4 DRAM SCALING TREND

3.4.1 SCALING CHALLENGES

Since its invention in the late 1960s, DRAM has witnessed an exponential scaling trend similar to the logic transistors predicted by Moore's law. Figure 3.17 shows the historical trend of DRAM's minimum feature size (F) scaling, partly driven by the advancements in lithography. Since the mid-2010s, DRAM has entered the sub-20 nm regime, and lithography is no longer the bottleneck of scaling. DRAM's major scaling challenges are summarized as follows: First, from the sense margin's perspective, maintaining sufficient SN capacitance is still the key. This is especially challenging when the parasitic BL capacitance does not scale noticeably for many generations. As a result, C_{SN} only changes from ~40 fF in the late 1980s to ~10 fF in

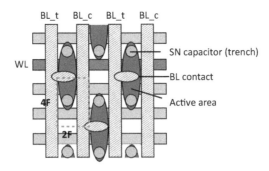

FIGURE 3.15 Layout of the trench-capacitor-based folded bit line architecture with 8 F^2 cell area.

Dynamic Random Access Memory (DRAM)

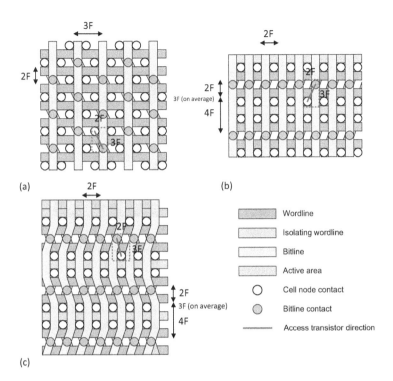

FIGURE 3.16 Layout variants of the stacked-capacitor-based open bit line architectures with 6 F^2 cell area with titled channel (active area).

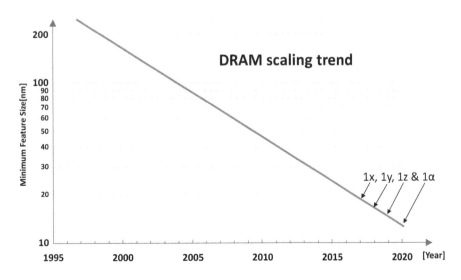

FIGURE 3.17 The historical trend of DRAM's minimum feature size (F) scaling.

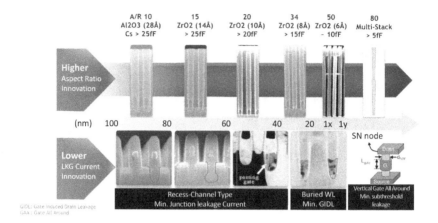

FIGURE 3.18 Technological innovations of the DRAM cell capacitor and access transistor technologies that enable continued scaling.

the late 2010s. Second, from the refresh's perspective, the increased leakage from the cell access transistor's short channel effect is a concern for data retention. Third, from the speed's perspective, the increased WL and BL interconnect resistivity due to the enhanced surface scattering along narrow wires is detrimental. There are many innovations in the past decades to overcome these challenges, as highlighted in Figure 3.18. On the cell capacitor side, higher and higher aspect ratios are used, as well as smaller and smaller equivalent oxide thicknesses (EOT) are enabled by high-k dielectric materials. On the cell access transistor side, new device structures to suppress leakage current are applied. Looking forward, nevertheless, DRAM will see fundamental roadblocks for scaling toward sub-10 nm generations.

3.4.2 Cell Capacitor

The cell capacitor determines the SN node capacitance. As indicated by the simple plate capacitor formula,

$$C = \varepsilon_0 \varepsilon_r A/t \tag{3.7}$$

where ε_0 is the vacuum permittivity constant, ε_r is the relative permittivity or the dielectric constant, A is the surface area of the electrode, t is the dielectric thickness. In a 3D vertical cylindrical capacitor structure, C_{SN} is given by

$$C_{SN} = \varepsilon_0 \varepsilon_r \times \pi \times AR \times F^2/t \tag{3.8}$$

where AR is the aspect ratio that is defined as the cylinder height divided by its diameter, F is the feature size or the critical dimension of the DRAM technology node. EOT could be introduced to normalize any dielectric to SiO_2 (whose dielectric constant is 3.9), since EOT $= \varepsilon_{SiO2}/\varepsilon_r \times t$; thus, C_{SN} is expressed as

$$C_{SN} = \varepsilon_0 \varepsilon_{SiO2} \pi \times AR \times F^2/EOT \tag{3.9}$$

As shown in Figure 3.18, the AR has been increased from 10 to more than 80. One potential problem of such extremely high AR is its mechanical stability. To avoid falling apart, mechanical support layers have been introduced in the sub-20 nm DRAM process, to have multi-stacking capability for a tall capacitor.

The evolution of the capacitor dielectric materials used in DRAM to understand the EOT scaling is discussed in the next. In retrospect, the silicon/insulator/silicon (SIS) structure was employed in the trench capacitor era, where the two electrodes were both heavily doped poly polysilicon and oxide/nitride/oxide (ONO) layered dielectric was used. Later toward the end of the trench capacitor era in the early 2000s, the metal/insulator/metal (MIM) structure was adopted. Starting from the stacked capacitor era, conductive TiN was used as a metal electrode and high-k dielectric materials such as Al_2O_3 were employed, which recently was switched to ZrO_2 (or its alloys with HfO_2). The dielectric constant of these optimized materials ranges from 20 to 30. As shown in Figure 3.18, EOT has been decreased from 2.8 nm to sub-0.6 nm (the Armstrong regime) thanks to the high-k dielectric material. Given the EOT = 0.6 nm, the high-k dielectric could relax the physical thickness to 3–5 nm; thus, the direct quantum tunneling current through the dielectric layer is much suppressed. It should be noted that a good SN capacitor should not only provide high capacitance, but also need to suppress leakage current for maintaining DRAM's retention. Using even higher-k material such as TiO_2 or $SrTiO_3$ is undesired as there is a well-known trade-off between the permittivity and the bandgap of the dielectrics. Higher permittivity means narrower bandgap, and thus exponentially larger leakage current. Figure 3.19 shows the recent trends of EOT and AR for DRAM capacitor's material and structure.

Another structural innovation that enables further scaling in the sub-20 nm DRAM generations is to switch the cylinder-type capacitor to the pillar-type capacitor, as shown in Figure 3.20. The critical dimension of the cylinder-type includes twice the thickness of the SN electrode layer, twice the thickness of the dielectric layer and the diameter of the inner plate electrode. The critical dimension of the pillar-type includes just twice the thickness of the dielectric layer and the diameter of the inner plate electrode. Therefore, the pillar-type saves some spacing occupied by the SN electrode layer. Instead, the SN electrode is fabricated as a standing-up pillar.

3.4.3 INTERCONNECT

The BL parasitic capacitance also plays an important role in determining the sense margin. Figure 3.21 shows the major components of the BL parasitic capacitance: (1) the coupling between WL and BL contact via; (2) the coupling between SN capacitor and BL; (3) the coupling between adjacent BLs; (4) the coupling between the BL and the substrate. The pie chart also suggests that the first two components dominate the total parasitic capacitance. Between these interconnect wires, isolation dielectric materials are filled. Using low-k dielectric (lower than SiO_2) is thus a natural way to reduce the coupling capacitances. The lowest material on the earth is the air (ε_r approaching 1). Therefore, an engineering solution to introduce the air spacer (or air gap) between interconnects has been introduced in sub-20 nm DRAM generations.

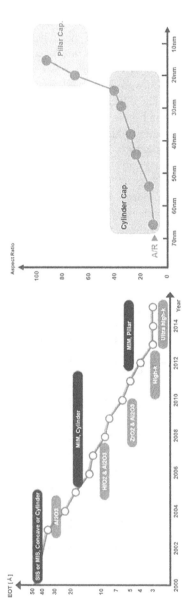

FIGURE 3.19 The recent trends of EOT and AR for DRAM capacitor's material and structure.

Dynamic Random Access Memory (DRAM)

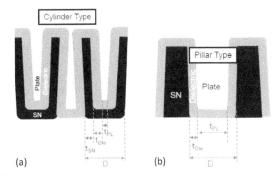

FIGURE 3.20 Stacked DRAM capacitor structures that change from (a) cylinder-type capacitor to (b) pillar-type capacitor.

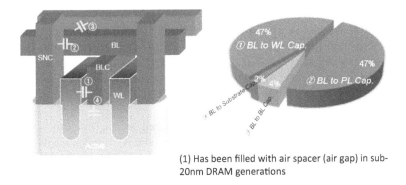

(1) Has been filled with air spacer (air gap) in sub-20nm DRAM generations

FIGURE 3.21 The major components of the BL parasitic capacitance and the breakdown of their contribution.

3.4.4 CELL ACCESS TRANSISTOR

The DRAM's cell access transistor that controls the access to the SN capacitor needs to be specially engineered to ensure ultra-low leakage current. Here is a rough estimation: for $C_{SN} = 10$ fF, and a refresh cycle = 64 ms, a degradation of voltage swing by 100 mV requires the leakage current $I_{leak} = 10$ fF $\times$ 0.1 V/64 ms = 15.6 fA. This is orders of magnitude lower than the low power logic transistor's off-state current (10–100 pA).

Figure 3.22 compares the differences between the logic transistor and DRAM cell access transistor. Logic transistor is designed for high-speed operations; thus, thin EOT gate oxide is used for obtaining large on-state current. DRAM cell access transistor is NMOS only, and requires high density to fit within the 6 F^2 compact design rule, and most importantly should offer ultra-low leakage. To overdrive with relatively higher V_{PP}, a thick EOT gate oxide is used; thus, the on-current is moderate. The most notable feature of DRAM cell access transistor is the recessed channel that is used to suppress the off-state current. To combat with the short-channel effect, it is intuitive to increase the effective channel length by the U-shape while shrinking the lateral distance between source and drain.

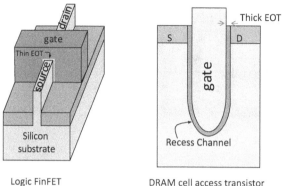

FIGURE 3.22 Comparison between logic transistor and DRAM cell access transistor.

In addition to the cell access transistor, DRAM peripheral transistor that is used for peripheral circuits is also different from the state-of-the-art logic transistor. DRAM peripheral transistor aims for low cost with barely sufficient performance, similar to some old logic process (e.g., 65 nm). Due to the dramatic differences in the fabrication processes in DRAM cell access transistor and the requirement of a high aspect ratio of cell capacitor, the DRAM manufacturing is usually done in a separate fabrication plant rather than in a logic foundry.

Figure 3.23 shows the milestones of DRAM cell access transistor scaling. At 90 nm node, the U-shape recess channel was introduced to suppress the off-state current. At 45 nm node, the FinFET was introduced to be combined with recess channel, namely saddle-FinFET, which offers better immunity to the short-channel

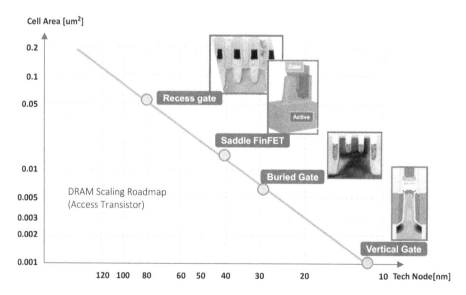

FIGURE 3.23 The trend of DRAM cell access transistor scaling.

Dynamic Random Access Memory (DRAM)

effect. At 32 nm node, the buried gate was introduced to reduce the coupling between WL to SN capacitor, as the gate is buried underneath the silicon surface.

Figure 3.24 summarizes state-of-the-art LPDDR products in the 1z node from primary vendors. F is around 16 nm, and the bit cell size is less than 0.002 μm². EUV is also introduced in DRAM manufacturing to further boost the bit density up to 0.27 Gb/mm², and it is expected that EUV will extend the DRAM's 2D scaling in the 2020s toward 10 nm node.

If carefully calibrating the recent scaling data in Figure 3.2, it is found that the DRAM cell has departed from the theoretical 6 F^2 for open bit line architecture in sub-20 nm generations. The WL pitch poses a constraint due to the seriousness of the short-channel effect at low voltage, as well as the row hammer effect due to WL coupling. This forces the active island to align more parallel with the BL. While this helps keep enough landing space for the stacked capacitor contact, the number of F^2 is going up in the latest nodes. For this reason, 4 F^2 design rule with vertical access transistor continues to appear attractive for the future scaling of DRAM beyond 10 nm node. Figure 3.25 shows the proposed 4 F^2 design rule with vertical access transistor that is still under research and development (as of 2020). The vertical channel is standing on top of the BL and is perpendicularly controlled by a gate-all-around WL. The stacked capacitor is integrated on top, allowing a theoretical 4 F^2 design rule.

Device	Micron D1z	Samsung D1z	Samsung D1z
Memory Capacity	16 Gb	12 Gb	16 Gb
Die Size	68.34 mm²	43.98 mm²	61.20 mm²
Bit Density (Die)	0.234 Gb/mm²	0.273 Gb/mm²	0.261 Gb/mm²
Cell Size	0.00204 μm²	0.00197 μm²	0.00197 μm²
Design rule	15.9 nm	15.7 nm	15.7 nm
EUV Lithography Applied	No	Yes	No

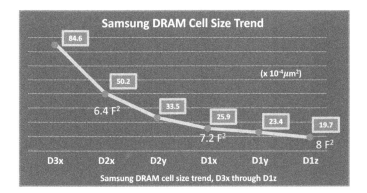

FIGURE 3.24 Summary of 1z node DRAM from major vendors and the trend of cell size that deviates from the 6 F^2 rule in the recent nodes.

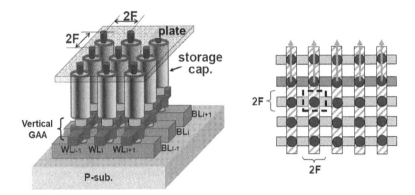

FIGURE 3.25 The proposed 4 F^2 design rule and layout with the gate-all-around (GAA) vertical access transistor for DRAM.

3.5 3D STACKED DRAM

3.5.1 TSV Technology and Heterogeneous Integration

As aforementioned, DRAM has faced fundamental challenges in 2D scaling due to the requirements of higher aspect ratio (that is lithography and etching demanding) and higher-k dielectric (while maintaining leakage current low). It is thus necessary to seek an alternative approach to improve the DRAM's performance at the system level. One promising approach is to explore the third dimension by staking multiple 2D DRAM dies to further increase the integration density and potentially boost the I/O bandwidth. The technology enablement for such 3D integration of multiple dies is the through-silicon-via (TSV). TSV is able to make a metallic conductive via through the silicon wafer (or die) that is thinned to several tens of μm to connect the front side interconnect and the back side interconnect. Figure 3.26 shows the typical fabrication process flow of the TSV: (1) deep trench is etched into silicon; (2) isolation dielectric is deposited covering the trench; (3) copper diffusion barrier and adhesion layer are deposited covering the trench; (4) copper is electroplated to fill in the trench; (5) silicon surface is planarized to remove residual copper by chemical-mechanical-polishing (CMP); (6) wafer/die thickness is shrunk by grinding. Figure 3.26(b) summarizes the typical TSV parameters. In general, the TSV pitch range is 10 μm–50 μm; the TSV diameters range is 5 μm–25 μm; the TSV aspect ratio range is 5–20; the TSV resistance range is 0.01–0.1 Ω; the TSV capacitance range is 50 fF–500 fF.

With the recent progress in TSV, the 2.5D and 3D heterogeneous integration have become mainstream technologies in the advanced packaging for DRAM. Figure 3.27(a) shows a 2.5D heterogeneous integration example: die 1 (e.g., a logic processor) and die 2 (e.g., a DRAM chip) are bumped on the silicon interposer (where some sort of horizontal interconnect could bridge die 1 and die 2). Then TSVs are built in the silicon interposer to allow the external I/O and power signals to route through the silicon interposer from the package substrate. Such 2.5D integration could be generalized for the so-called "chiplet" integration, with various functional

Dynamic Random Access Memory (DRAM)

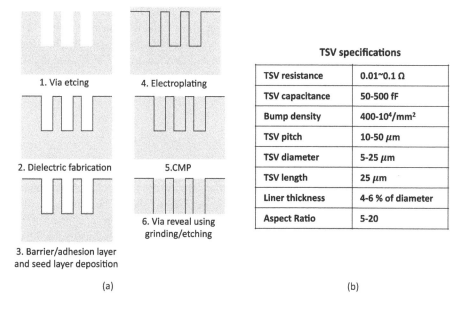

FIGURE 3.26 (a) Fabrication process flow of the TSV. (b) Typical TSV parameters.

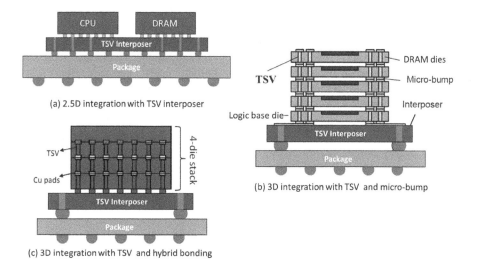

FIGURE 3.27 (a) 2.5D heterogeneous integration; (b) 3D heterogeneous integration with TSV and micro-bump; (c) heterogeneous integration with bumpless hybrid bonding.

dies (CPU, GPU, DRAM, Flash, image sensors, MEMS, RF transceivers, etc.[3]) being packaged on the same interposer for building system-in-package. Figure 3.27(b) shows a 3D heterogeneous integration example: multiple dies are vertically stacked using the TSVs and micro-bumps. Micro-bump with diameter around 30 μm is a key enabler for such 3D heterogeneous integration that has been widely used in 3D stacking DRAM dies on top of a logic base die. Figure 3.27(c) shows an example of the next-generation 3D heterogeneous integration, namely hybrid bonding. The hybrid bonding is bumpless and copper pads are directly bonded together. Therefore, the pad pitch (or the TSV pitch) could be reduced to a few μm. In the future, hybrid bonding with nano-TSV aim for sub-μm pitch is feasible as demonstrated in recent research [3]. Pitch scaling allows higher connectivity between dies and results in larger bandwidth.

3.5.2 HBM

High-bandwidth memory (HBM) is a 3D stacked DRAM system using the TSV and micro-bumps. HBM provides the potential solutions to the DRAM performance boost beyond the traditional 2D scaling. Figure 3.28(a) shows the schematic of a high-performance computing platform that uses 2.5D integration to connect a 3D HBM stack with a GPU. Multiple DRAM dies are vertically stacked using the TSVs and micro-bumps on top of a logic base die that serves as the memory controller. The logic base die communicates with the GPU horizontally on the silicon interposer that sits on a package substrate. Figure 3.28(b) shows the top-view of a HBM die. TSV array is placed in the middle of the die.

What makes HBM attractive is not only the higher integration density (multiple dies on the same 2D form factor), but also the wide I/O interface that it could offer. As aforementioned, the DDR/LPDDR often has a 64-bit-wide I/O interface, and GDDR often has a 32-bit-wide I/O interface. Now, HBM offers a 1024-bit-wide I/O interface. Even running at a slower I/O clock frequency, which means lower interface speed (Gbps) per pin, HBM could offer a significantly higher bandwidth (GB/s) at the system level. Table 3.1 summarizes the evolution of HBM interface protocol standard and the comparison with LPDDR and GDDR counterparts. As of 2020, HBM has gone through three generations (HBM, HBM2, and HBM2E). The capacity per DRAM die has increased from 2 Gb to 16 Gb, and the number of DRAM dies in the stack has increased from 4 to 8; thus, the total capacity has increased from 1 GB to 16 GB. The system bandwidth has increased from 128 GB/s to 410 GB/s. As a comparison, LPDDR5 and GDDR6 offer the system bandwidth 37.5 GB/s and 56 GB/s.

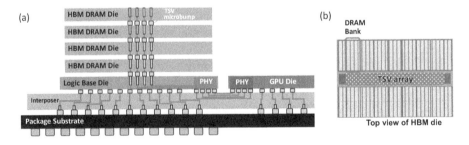

FIGURE 3.28 (a) Schematic of a high-performance computing platform that uses 2.5D integration to connect a 3D HBM stack with a GPU. (b) Top view of an HBM die.

Dynamic Random Access Memory (DRAM)

TABLE 3.1
The Evolution of HBM Interface Protocol Standard, and the Comparison with LPDDR and GDDR Counterparts

Parameters	DDR5	LPDDR5/4	GDDR6	HBM1	HBM2	HBM2E
Interface Speed (Gbps) per pin	4.2	4.2	14	1	2.4	3.2
Density	8 Gb	8 Gb	8 Gb & 16 Gb	2 Gb	8 Gb	16 Gb
Bus Width	64-bit	64-bit	32-bit	1024-bit	1024-bit	1024-bit
Bandwidth	33.6 GB/s	33.6 GB/s	56 GB/s	128 GB/s	307 GB/s	410 GB/s
Applications	Consumer	Mobile consumer	GPU, Automotive	HPC, AI, Networking	HPC, AI, Networking	HPC, AI, Networking

Figure 3.29 shows a system-level performance comparison between a platform with GDDR5 and a platform with HBM. The HBM platform not only reduces the form factor (package area) by 75% and boosts the bandwidth by 3.6×. These different DRAM options with different interfaces have different targeted applications. DDR family aims for low-cost personal computer. LPDDR family aims for low-power mobile platforms. GDDR family aims for GPU and automotive. HBM family aims for high-performance computing including GPU, networks of accelerators for AI workloads.

ITEM		GDDR5	HBM
System 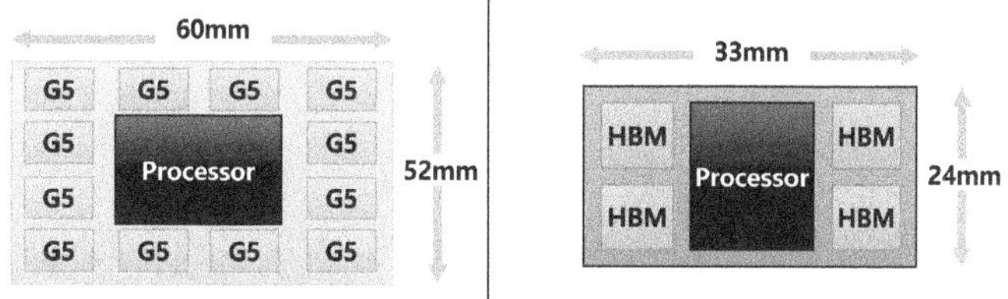		60mm × 52mm (Processor surrounded by 12 G5 chips)	33mm × 24mm (Processor with 4 HBM)
DRAM		8Gb GDDR5 12ea	4GB HBM 4ea
Size		3120 mm² (−75%)	792 mm²
Density		12GB (1.3×)	16GB
Bandwidth		384GB/s (3.6×)	1024GB/s
Power		18.3W (1.5W × GDDR5 12ea) (+18%)	9.1W (2.3W × HBM 4ea)
Pin (Ball)	Speed	8 Gbps	2 Gbps
	# I/O	32 per chip (Total 384)	1024 per cube (Total 4096)

FIGURE 3.29 System-level performance comparison between a platform with GDDR5 and a platform with HBM.

3.6 EMBEDDED DRAM

3.6.1 1T1C eDRAM

Embedded DRAM (eDRAM) is a variant of DRAM that is integrated on the same chip of the logic process primarily for the last-level cache with ultra-large capacity (100 MB–1 GB). Unlike standalone DRAM that uses separate fabrication processes, eDRAM is fully compatible to the logic process. eDRAM is positioned somewhere between SRAM and standalone DRAM in terms of the cell area (or the capacity) and the access speed. As aforementioned, SRAM has a typical cell area ranging from 150 F^2 to 300 F^2, and standalone DRAM has a typical cell area of 6 F^2, while eDRAM has a typical cell area ranging from 30 F^2 to 90 F^2. SRAM can be accessed within 1 ns, standalone DRAM can be accessed within several tens of ns (row cycle time), while eDRAM can be accessed within a few ns. However, a notable drawback of eDRAM is its reduced retention to 100 μs due to increased leakage current of cell access transistor in the logic process.

The 1T1C-based eDRAM has been developed by two major companies. The first one is IBM (and its foundry partner Globalfoundries). IBM has been using eDRAM for many generations for their POWER series processor for high-performance computing. IBM uses a SOI technology for the logic process, and uses a deep trench capacitor for the eDRAM. In this scenario, the eDRAM cell is fabricated first and then the logic transistor is processed. Figure 3.30 shows the scaling trend of IBM's eDRAM from 65 nm to 14 nm, which tends to slow down, resulting in larger cell area in F^2 if normalized to the technology node. Power 8 processor introduced in 2014 is equipped with 96 MB eDRAM L3 cache at 22 nm planar SOI platform [4]. Power 9 processor introduced in 2017 is equipped with 120 MB eDRAM L3 cache at 14 nm FinFET SOI platform [5]. The second one is Intel, who developed eDRAM at its 22 nm FinFET bulk platform in 2014 [6, 7], as shown in Figure 3.31. Unlike IBM, Intel's approach is using the stacked capacitor with a moderate aspect ratio of about 8. Though the research results appear promising, there has been no commercial product from Intel that uses eDRAM yet.

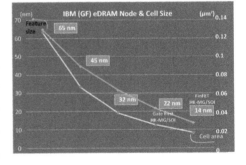

eDRAM Technology	IBM Power8™ 22nm Planar HK-MG SOI	IBM Power9™ 14 nm HP FinFET HK-MG SOI
eDRAM Cell Size	0.026 um²	0.0174 um²
Normalized Cell Area	54F²	89F²
SN Capacitance (estimated)	~12.2 fF	~8.1 fF
L3 cache eDRAM Density	11.9 Mb/mm²	13.28 Mb/mm²

FIGURE 3.30 IBM's deep-trench-based eDRAM parameters in Power 8 and Power 9 processors and the scaling trend from 65 nm to 14 nm.

Dynamic Random Access Memory (DRAM)

Technology	22nm FinFET (Intel)	
Cell Size	$0.029 \mu m^2$	
Supply	1.05V	
	1st–Gen eDRAM	2nd–Gen eDRAM
Clock, Random Cycle Time	2GHz, 3ns	2GHz, 5ns
Retention Time	100μs @ 95°C	300μs @ 95°C

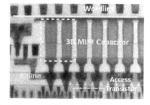

Capacitor over Bit-line (COB) architecture

FIGURE 3.31 Intel's eDRAM demonstration in 22 nm node with stacked capacitor over bit line.

3.6.2 Capacitor-Less eDRAM

Alternative designs that are capacitor-less have been proposed to build eDRAM. Getting rid of the cell capacitor helps to scale the bit cell area. One feasible design is a 1-transistor floating body cell based on SOI technology, as shown in Figure 3.32. With a negative gate-to-drain voltage, the GIDL effect creates the electron/hole pairs due to the band-to-band tunneling. As electrons are collected to the drain, the holes are supposed to sink to the body in the bulk transistor. Nevertheless, the SOI transistor does not have a body terminal and the body is essentially floating due to the buried oxide (BOX) layer in the substrate. Therefore, the holes are accumulated in the floating body, which effectively lowers down the source to the channel barrier, thus leading to a higher drain current. In other words, the threshold voltage is lowered if there is hole accumulation in the floating body. Figure 3.32 also shows the $I_D - V_G$ transfer curve for the two states (hole accumulation for "1", and no hole accumulation for "0"). The read-out is to measure the drain current difference, and this is a non-destructive process. However, the hole accumulation is not forever, as it may recombine with

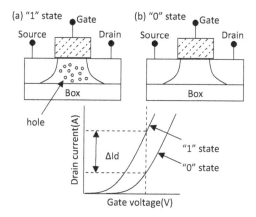

FIGURE 3.32 Principle of the capacitor-less eDRAM based on 1-transistor floating body cell with SOI technology.

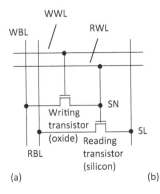

Gain Cell Type	2T (SI)	2T (IGZO)	2T (IWO)
V_{DD}(V)	1.2	1.2	1.0
Density (Mb/mm²)	4	80	180
BEOL Compatible for 3D Stacking	No	Yes	Yes
Cell Cap. (fF/cell)	N/A	1.2	1
Retention (ms@85°C)	0.1	10^7	10^3
Access Time (ns)	1.6	30	3
Destructive Read	No	No	No
Channel Material of Reading/Writing Transistor	Si/Si	IGZO/IGZO	Si/IWO

FIGURE 3.33 (a) Schematic of 2T gain cell capacitor-less eDRAM with oxide channel transistor as the write transistor to minimize the leakage current. (b) Summary of the representative 2T gain cell design reported in the literature.

minority electrons in the body. Therefore, a refresh is still needed. Advanced band engineering is necessary to boost the floating-body-based eDRAM [8].

Another feasible design is to use the 2T gain cell, as shown in Figure 3.33(a). There are two transistors in one bit cell: the reading transistor and the writing transistor. The gate capacitor of the reading transistor becomes the storage node, which could be charged or discharged by the writing transistor. The stored electrons on the gate capacitor could suppress the channel current of the reading transistor, as it equivalently increases the threshold voltage. A small amount of charges on the gate could result in significant changes of the channel current as a result of the transconductance amplification. However, the gate capacitor has a limited capacitance thus cannot hold the charges for a long time (typically less than 1 ms if implemented with silicon logic process [9]). Therefore, the writing transistor should be ultra-low leakage. One promising candidate for the writing transistor is the semiconducting oxide-channel-based transistor, which could offer ultra-low off-state current density (<1 fA/μm) thanks to its wide bandgap. Recent progresses in In-Ga-Zn-Oxide (IGZO) [10] and In-W-Oxide (IWO) [11] offers the unique advantages as these oxide-channel-based transistors could be fabricated on top of the logic transistor at the BEOL, and thus no significant area penalty with 3D folded layout. Figure 3.33(b) summarizes the representative 2T gain cell design, highlighting the orders of magnitude improved retention by using oxide channel for the writing transistor's channel. It should be noted that silicon is still preferred for the reading transistor's channel for faster access as Si still outperforms the oxides by 5× to 10× in terms of electron mobility.

While eDRAM technology is not widely used in today's processor technologies due to issues such as process complexity and fabrication cost, there is a great interest to develop the next-generation eDRAM to satisfy the ever-growing demands of larger on-chip embedded memories, especially for the hardware accelerators for the data-intensive AI applications.

NOTES

1. Positive charge flow in fact means negative electron flow in the reverse direction.
2. Below 20 nm node, DRAM vendors no longer show the exact critical dimension, and just use the 1x, 1y, 1z, 1α symbols to represent different generations of the technology. 1α node has F = 14 nm approximately.
3. MEMS is micro-electromechanical system, and RF is radio frequency.

REFERENCES

[1] JEDEC, https://www.jedec.org/
[2] O. Mutlu, J.S. Kim, "RowHammer: a retrospective," *IEEE Transactions on Computer-Aided Design of Integrated Circuits and Systems*, vol. 39, no. 8, pp. 1555–1571, August 2020. doi: 10.1109/TCAD.2019.2915318
[3] E. Beyne, S.-W. Kim, L. Peng, N. Heylen, J.D. Messemaeker, O.O. Okudur, A. Phommahaxay, et al., "Scalable, sub 2µm pitch, Cu/SiCN to Cu/SiCN hybrid wafer-to-wafer bonding technology," *IEEE International Electron Devices Meeting (IEDM)*, 2017, pp. 32.4.1–32.4.4. doi: 10.1109/IEDM.2017.8268486
[4] E.J. Fluhr, J. Friedrich, D. Dreps, V. Zyuban, G. Still, C. Gonzalez, A. Hall, et al., "POWER8™: a 12-core server-class processor in 22nm SOI with 7.6Tb/s off-chip bandwidth," *IEEE International Solid-State Circuits Conference Digest of Technical Papers (ISSCC)*, 2014, pp. 96–97. doi: 10.1109/ISSCC.2014.6757353
[5] C. Gonzalez, E. Fluhr, D. Dreps, D. Hogenmiller, R. Rao, J. Paredes, M. Floyd, et al., "POWER9™: a processor family optimized for cognitive computing with 25Gb/s accelerator links and 16Gb/s PCIe Gen4," *IEEE International Solid-State Circuits Conference (ISSCC)*, 2017, pp. 50–51. doi: 10.1109/ISSCC.2017.7870255
[6] F. Hamzaoglu, U. Arslan, N. Bisnik, S. Ghosh, M.B. Lal, N. Lindert, M. Meterelliyoz, et al., "A 1Gb 2GHz embedded DRAM in 22nm tri-gate CMOS technology," *IEEE International Solid-State Circuits Conference (ISSCC)*, 2014, pp. 230–231. doi: 10.1109/ISSCC.2014.6757412
[7] M. Meterelliyoz, F.H. Alamoody, U. Arslan, F. Hamzaoglu, L. Hood, M. Lal, J.L. Miller, et al., "2nd generation embedded DRAM with 4X lower self refresh power in 22nm tri-gate CMOS technology," *IEEE Symposium on VLSI Circuits*, 2014, pp. 1–2. doi: 10.1109/VLSIC.2014.6858415
[8] C. Navarro, S. Karg, C. Marquez, S. Navarro, C. Convertino, C. Zota, L. Czornomaz, F. Gamiz. "Capacitor-less dynamic random access memory based on a III–V transistor with a gate length of 14 nm," *Nature Electronics*, vol. 2, no. 9, pp. 412–419, 2019. doi: 10.1038/s41928-019-0282-6
[9] K.C. Chun, P. Jain, T. Kim, C.H. Kim, "A 1.1V, 667MHz random cycle, asymmetric 2T gain cell embedded DRAM with a 99.9 percentile retention time of 110µsec," *IEEE Symposium on VLSI Circuits*, 2010, pp. 191–192. doi: 10.1109/VLSIC.2010.5560303
[10] M. Oota, Y. Ando, K. Tsuda, T. Koshida, S. Oshita, A. Suzuki, K. Fukushima, et al., "3D-stacked CAAC-In-Ga-Zn oxide FETs with gate length of 72nm," *IEEE International Electron Devices Meeting (IEDM)*, 2019, pp. 3.2.1–3.2.4. doi: 10.1109/IEDM19573.2019.8993506
[11] H. Ye, J. Gomez, W. Chakraborty, S. Spetalnick, S. Dutta, K. Ni, A. Raychowdhury, S. Datta, "Double-gate W-doped amorphous indium oxide transistors for monolithic 3D capacitorless gain cell eDRAM," *IEEE International Electron Devices Meeting (IEDM)*, 2020, pp. 28.3.1–28.3.4. doi: 10.1109/IEDM13553.2020.9371981

4 Flash Memory

4.1 FLASH OVERVIEW

4.1.1 Flash's History

The simplest form of digital data storage in semiconductors is read-only memory (ROM), which strictly refers to the memory that is hard-wired, such as a mask-based interconnect matrix where inputs and outputs are short-circuited ("1") or open-circuited ("0"), as defined by photolithography, which cannot be electrically changed after manufacturing. The programmable ROM (PROM) that allows users to write the data once is called the one-time programmable (OTP) memory. One example is the antifuse-based OTP with dielectric or amorphous silicon thin film sandwiched between two metal electrodes. Initially, with an insulating state ("0"), a programming pulse with high voltage breaks down the thin film and converts it into a permanent conductive path ("1"). This process is irreversible. OTP is still used today in the boot code, encryption keys, and firmware parameters for analog, sensor, or display circuitry.

The Flash's history could be traced back to the erasable PROM (EPROM), where data could be written multiple times. EPROM was based on the floating-gate transistor. In 1967, Dawon Kahng and Simon Min Sze at Bell Labs proposed the concept of a floating-gate transistor. Building on this concept, Dov Frohman of Intel invented the commercial EPROM in 1971. The program of EPROM (write "0") is based on the channel hot electrons (CHEs) that are trapped in the floating gate, however, the erase of EPROM (write "1") could not be done electrically. Instead, ultraviolet light was used to excite electrons out from the floating gate. Later in the 1970s, an improved version of EPROM was invented to allow it to be electrically erasable and programmable, namely EEPROM (also called E^{2}PROM). The erase operation is done by applying high voltage into the source, and the electrons are removed from the floating gate by the Fowler–Nordheim (F-N) tunneling. EEPROM's mechanism is very similar to today's Flash memory (especially the NOR Flash). EEPROM typically employed a 2-transistor bit cell, as shown in Figure 4.1. The floating-gate transistor is selected by a normal access transistor, which enables bit-addressable erase (i.e., erase a random bit by the given address).

Modern Flash memory was then introduced in the 1980s. Fujio Masuoka from Toshiba proposed a variant of EEPROM that allowed an entire section of memory to be erased together quickly and efficiently, by applying a voltage to a single wire connected to a group of cells. The name "Flash" memory came from the analogy: the fast erase process of the memory contents versus the Flash of a camera. Masuoka and colleagues from Toshiba presented the invention of NOR Flash in 1984 at the

DOI: 10.1201/9781003138747-4

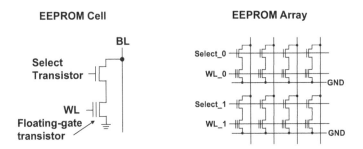

FIGURE 4.1 Schematic of EEPROM cell and array. The bit cell consists of a select transistor and a floating-gate transistor.

International Electron Devices Meeting (IEDM) [1], and then NAND Flash in 1987 at the IEDM [2].

Compared to EEPROM, Flash memory has only one floating-gate transistor per bit thus is designed for high density, which also offers faster erase throughput with relatively large erase sections simultaneously (typically 512 Bytes or larger). There is no clear boundary dividing the two, but the term EEPROM is generally reserved to describe the 2-transistor bit cell with bit-addressable erase capability. EEPROM is still being used today in microcontrollers (e.g., for smart cards or remote keyless systems) to store relatively small amounts of data (tens to hundreds kB) and it offers reasonably long program/erase endurance ($>10^6$ cycles).

4.1.2 Flash's Applications

Since its invention in the 1980s, Flash memory (especially the NAND Flash) has become the technological foundations of the following digital storage products: (1) SD memory card; (2) USB memory stick; (3) solid-state drive (SSD). The landscape of Flash-based products is shown in Figure 4.2. SD, standing for Secure Digital, is a memory card format developed for use in portable devices (e.g., mobile phones and tablets, digital cameras, etc.). The standard was introduced in 1999 by joint efforts between SanDisk, Panasonic, and Toshiba. The memory card has offered different form factors (from a larger size to a smaller size) including SD and microSD. The SD capacity standard has extended a few generations so far, including SD (up to 2 GB), SDHC (up to 32 GB), SDXC (up to 2 TB), and SDUC (up to 128 TB). The recent SD Express interface protocol could enable 1 GB/s to 4 GB/s transfer rate from SD cards. The USB memory stick is a Flash drive with an integrated Universal Serial Bus interface. The USB interface protocol has evolved generations from 1.0 to 2.0, 3.0, and 4.0 with the transfer rate improved from 1.5 MB/s to 60 MB/s, 625 MB/s, and 5 GB/s. The typical USB Flash drive offers the capacity of 32 GB up to 2 TB as of 2020, and it is primarily used for digital media storage and transfer between devices. SSD is the massive data storage technology that aims to replace the hard-disk drive (HDD). As of 2020, SSD was widely available in the capacity range of 128 GB–4 TB for consumer electronics, and up

Flash Memory

FIGURE 4.2 The data storage products that are enabled by Flash memory technology.

to 100 TB for enterprise electronics. On the consumer side, SSD has become a mainstream platform for personal computers and laptops. On the enterprise side, SSD is emerging as the driving force for innovating data centers. NVMe, short for Non-Volatile Memory Express, is an interface protocol built especially for SSD. NVMe works with Peripheral Component Interconnect Express (PCIe) to transfer data to and from SSD. NVMe is an improvement over the older HDD-related interfaces such as Serial AT Attachment (SATA). For example, the SATA III protocol maxes out at a throughput of 600 MB/s for SSD and 100 MB/s for a 7200 rotations per minute (RPM) HDD. NVMe drives, on the other hand, provide transfer rates as high as 3500 MB/s or even higher. The reason why SATA is sometimes used for SSD in personal computers is that the back-compatibility with the HDD systems.

4.2 FLASH DEVICE PHYSICS

4.2.1 Principle of Floating-Gate Transistor

The floating-gate transistor has been widely used in NOR Flash and 2D NAND Flash. Figure 4.3(a) shows the schematic of the floating-gate transistor's structure, which is generally using a p-type substrate (working as NMOS). Compared to the regular planar MOSFET, the unique feature of the floating-gate transistor is its additional floating gate, which is embedded between the control gate and the channel. The floating gate is essentially floating as it has no direct connection to external terminals. Between the floating gate and the channel is the tunnel oxide (e.g., 9-nm-thick SiO_2 layer), and between the floating gate and the control gate is the inter-poly dielectric (e.g., 15-nm-thick $SiO_2/Si_3N_4/SiO_2$, ONO layer). The floating gate and the control gate are typically doped poly-silicon. The tunnel oxide and inter-poly dielectric of the floating-gate transistor are much thicker than that of the regular logic transistor because a much higher voltage (e.g., 20 V) will be applied to the control gate.

Figure 4.3(b) shows the operational principle of the floating-gate transistor, its I_D–V_G (with respect to the control gate) transfer curve, and its circuit symbol. The program operation injects electrons into the floating gate, and then the negatively charged electrons partially screen the positive voltage applied on the control gate;

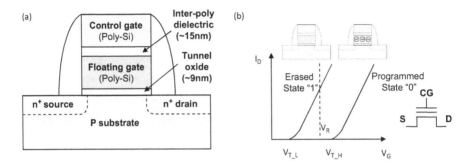

FIGURE 4.3 (a) Schematic of the floating-gate transistor's structure. (b) The operational principle of the floating-gate transistor.

thus, the threshold voltage is increased to V_{T_H} as it is more difficult to turn on the channel. The erase operation ejects electrons out from the floating gate, thus the threshold voltage is decreased to V_{T_L}.[1] When reading out the memory state, a read voltage V_R that is in between V_{T_H} and V_{T_L} is asserted to the control gate. Then, the erased state will be read out with a noticeable drain current (interpreted as "1"), while the programmed state will be read out with a negligible drain current (interpreted as "0"). Essentially, the memory state is stored as the charges (in this case, electrons) on the floating gate. The charges, once injected into the floating gate, should stay there for a long time even when the power supply is removed. Therefore, the floating-gate transistor is a type of non-volatile memory (NVM).

4.2.2 Capacitor Model for Floating-Gate Transistor

Electrostatically, a first-order capacitor model is presented to describe the threshold voltage memory window, and the floating gate voltage (V_{FG}) dependence on the external control gate (V_{CG}) voltage and the drain voltage (V_D), as shown in Figure 4.4. The capacitances are labeled as follows: C_g is the tunnel oxide capacitance, C_s is the coupling capacitance between the floating gate and the source induced by the fringing electric field, C_d is the coupling capacitance between the floating gate and the source

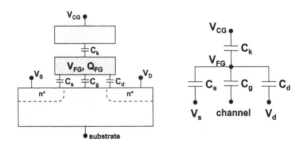

FIGURE 4.4 A simple capacitor model for the floating-gate transistor's electrostatics.

induced by the fringing electric field, and C_k is the inter-poly dielectric capacitance. Assuming that $V_S = V_D = V_B = 0$ while V_{CG} is applied, the stored charges (Q_{FG}) after the program is contributed by all the capacitances multiplied by the voltage across the capacitors that are connected to the floating gate which has a voltage potential V_{FG}:

$$Q_{FG} = (V_{FG} - V_{CG})C_k + V_{FG}C_s + V_{FG}C_g + V_{FG}C_d$$
$$= V_{FG}(C_k + C_s + C_g + C_d) - V_{CG}C_k = V_{FG}C_t - V_{CG}C_k \quad (4.1)$$

Here the total capacitance $C_t = C_k + C_s + C_g + C_d$. The V_{FG} could be expressed as

$$V_{FG} = C_k/C_t \, V_{CG} + Q_{FG}/C_t \quad (4.2)$$

To achieve the same drain current level, the floating-gate transistor should have the same V_{FG} as V_{FG} determines the channel surface potential. Therefore, the memory window is defined as the threshold voltage difference (ΔV_T) that is caused by the storage charges (Q_{FG}) on the floating gate. When there are no charges, $Q_{FG} = 0$ after erase, a different V_{CG}' could be applied; thus, Equation (4.2) becomes,

$$V_{FG} = C_k/C_t \, V_{CG}' \quad (4.3)$$

By equalizing Equation (4.2) and (4.3), ΔV_T could be obtained as follows:

$$\Delta V_T = V_{CG} - V_{CG}' = Q_{FG}/C_k \quad (4.4)$$

From Equation (4.3), the control gate to the floating gate coupling ratio α_{CG} could be defined as

$$\alpha_{CG} = C_k/C_t \quad (4.5)$$

Essentially α_{CG} determines how effectively the externally applied control voltage could modulate the internal floating gate voltage.

An experimental procedure is shown in Figure 4.5 to illustrate how to determine α_{CG} from the I_D–V_G transfer curve. A dummy cell is used, which is a regular MOSFET that has the same dimensions as the floating-gate transistor but without the floating gate. First, the floating-gate transistor is fully erased (thus Q_{FG} is zero) and it is biased with a small V_{DS} (e.g., 0.1 V). On ignoring the small V_{DS} and considering the relative changes to the voltages, Equation (4.3) could be expressed as

$$\Delta V_{FG} = \alpha_{CG} \, \Delta V_{CG} \quad (4.6)$$

On the other hand, for a dummy cell without a floating gate, the following equation holds:

$$\Delta V_{FG} = \Delta V_{G(\text{dummy})} \quad (4.7)$$

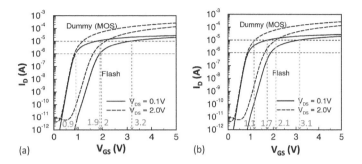

FIGURE 4.5 Example of I_D–V_G curve of a Flash transistor and a dummy MOSFET (without the floating gate). The information from these curves could be used to determine the coupling ratio α_{CG} from (a) and α_D from (b).

Therefore, one could measure ΔV_{CG} for floating-gate transistor and ΔV_G for dummy cell using I_D–V_G transfer curve by calculating the voltage differences between two current levels (e.g., 10^{-5} A and 10^{-6} A). Reading from Figure 4.5(a) gives

$$\alpha_{CG} = \frac{\Delta V_{G(\text{dummy})}}{\Delta V_{CG}} = \frac{V_G\left(I_D = 10^{-5}\right) - V_G\left(I_D = 10^{-6}\right)_{\text{dummy}}}{V_G\left(I_D = 10^{-5}\right) - V_G\left(I_D = 10^{-6}\right)_{\text{Flash}}} = \frac{1.9 - 0.9}{3.2 - 2} = 0.83 \quad (4.8)$$

Then, a large drain voltage (e.g., $V_D = 2$ V) is biased to determine the coupling ratio α_D between the drain voltage to the floating gate voltage. Now the floating voltage is also modulated by the drain voltage as follows:

$$\Delta V_{FG} = \alpha_{CG} \Delta V_{CG} + \alpha_D \Delta V_D \quad (4.9)$$

Given the known α_{CG} determined earlier in Equation (4.8), α_D could be obtained by comparing voltage differences between the small $V_D = 0.1$ V and large $V_D = 2$ V at a given current level (e.g., 10^{-5} A) in Figure 4.5(b),

$$\alpha_D = \frac{\Delta V_{FG} - \alpha_{CG} \Delta V_{CG}}{\Delta V_D} = \frac{\Delta V_{G(\text{dummy})} - \alpha_{CG} \Delta V_{CG}}{\Delta V_D}$$

$$= \frac{\left[V_G\left(V_D = 2V\right) - V_G\left(V_D = 0.1V\right)\right]_{\text{dummy}} - \alpha_{CG}\left[V_{CG}\left(V_D = 2V\right) - V_{CG}\left(V_D = 0.1V\right)\right]}{\left(V_D = 2V\right) - \left(V_D = 0.1V\right)}$$

$$= \frac{(1.1 - 1.7) - (2.1 - 3.1)}{2 - 1} = 0.12 \quad (4.10)$$

Ideally, a large α_{CG} (close to 1) is preferred since the control voltage could effectively modulate the floating gate voltage, and a small α_D (close to 0) is preferred since the drain voltage should not affect the floating gate voltage.

Flash Memory

4.2.3 Program/Erase Mechanism

The program/erase operations involve electrons injection/ejection into or out from the floating gate. Figure 4.6 summarizes the commonly used mechanisms. Figure 4.6(a) shows the CHE mechanism for the program, where a high voltage (e.g., $V_{CG} = 12$ V) is applied to the control gate, meanwhile, a relatively large voltage (e.g., $V_D = 6$ V) is applied to the drain. Under such biases, a portion of electrons flowing in the channel may gain enough kinetic energy to thermally jump over the tunnel oxide barrier into the floating gate. Figure 4.6(b) shows the source F-N tunneling mechanism for erase,[2] where a high voltage (e.g., $V_S = 15$ V) is applied to the source, meanwhile, the control gate is grounded (e.g., $V_{CG} = 0$ V). The electrons near the source may tunnel through the tunnel oxide barrier from the floating gate to the source in their overlap region. Figure 4.6(c) shows the channel F-N tunneling mechanism for the program, where a very high voltage (e.g., $V_{CG} = 20$ V) is applied to the control gate, meanwhile, the body/source/drain are all grounded. The electrons could tunnel through the tunnel oxide barrier in the entire channel region from the substrate to the floating gate. Figure 4.6(d) shows the channel F-N tunneling mechanism for the erase, where a very high voltage (e.g., $V_{sub} = 20$ V) is applied to the substrate, meanwhile, the gate/source/drain are all grounded. The electrons could tunnel through the tunnel oxide barrier in the entire channel region from the floating gate to the substrate.

Next, the F-N tunneling physics is discussed in more detail. Tunneling is a quantum mechanical phenomenon describing that the electrons may have a finite probability to pass through the energy potential barrier (e.g., SiO_2) if the barrier is thin

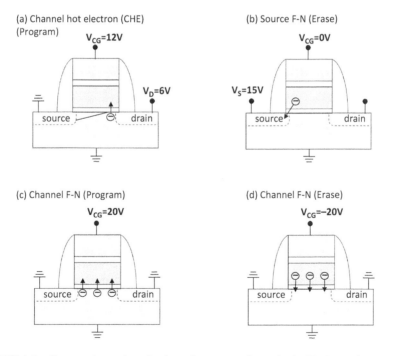

FIGURE 4.6 Commonly used mechanisms for program/erase in the Flash transistor.

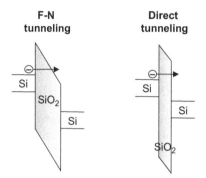

FIGURE 4.7 Energy band diagram for an Si/SiO$_2$/Si stack showing the F-N tunneling (with triangular barrier) and direct tunneling (with trapezoidal barrier).

(typically sub-10 nm). Depending on the shape of the barrier (trapezoidal or triangular), the tunneling may be direct tunneling or F-N tunneling for an Si/SiO$_2$/Si stack, as shown in Figure 4.7. The direct tunneling typically occurs in an ultra-thin (e.g., sub-2 nm) barrier. Take the channel F-N-tunneling-based program as an example, the typical tunnel oxide barrier is 8–10 nm and the floating-gate voltage is 16–20 V, resulting in the triangular shape of the barrier. The tunnel oxide barrier is designed so thick that the direct tunneling is prohibited in order to maintain sufficient retention time. The F-N tunneling follows the equation

$$J_{FN} = \alpha E_{ox}^2 \exp\left(-\frac{\beta}{E_{ox}}\right)$$

$$\text{where } \alpha = \frac{q^3 m_0}{16\pi^2 \hbar \Phi_B m^*}; \quad \beta = \frac{4\sqrt{2m^*}\Phi_B^{3/2}}{3\hbar q} \tag{4.11}$$

Here, J_{FN} is the current density, E_{ox} is the electric field in the oxide, Φ_B is the barrier height of the injecting interface, $\hbar$ is the reduced Planck constant, m_0 is the electron mass in vacuum, m^* is the effective electron mass, q is the basic charge quantity. It is noted that for an Si/SiO$_2$/Si stack, electrons Φ_B = 3.1 eV, and holes Φ_B = 3.8 eV. That is why in the floating-gate transistor, the electrons are easier to tunnel through and the stored charges are primarily electrons (instead of holes). For program/erase operations, F-N tunneling current is very small (typically < pA) such that the speed is relatively low (approximately 100 μs to 1 ms). However, the programming efficiency is high (approaching 100%), as all the electrons involved in the F-N tunneling contribute to the program/erase. Experimentally, the relationship between current density (J) through the gate stack and electric field (E) across the gate stack could be measured and $log(J)$ versus E is shown in Figure 4.8(a). The current density plot could be rearranged in the form of $log(J/E^2)$ versus $1/E$ as shown in Figure 4.8(b), which indicates a straight line for F-N tunneling and β could be extracted as the slope.

Flash Memory

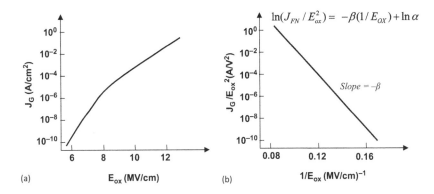

FIGURE 4.8 The relationship between current density (J) through the gate stack and electric field (E) across the gate stack. A straight line in log(J/E^2) vs. $1/E$ indicates the F-N tunneling mechanism dominates.

Next, the CHE effect is discussed in more detail. The floating-gate transistor is fully turned on (e.g., by $V_{CG} = 12$ V) and is in the saturation region (e.g., by $V_D = 6$ V); thus, electrons are accelerated by the lateral electric field as they travel along the channel from the source to the drain, causing the impact ionization process. Figure 4.9 shows the energy band diagrams along the channel direction (A-A') and perpendicular to the channel direction (B-B'). In the A-A' diagram, the band is significantly bent by the large drain voltage, the electrons that are approaching the drain will have high kinetic energy (indicated by the distance between the particle and the valence band edge). The kinetic energy is related to the temperature, and that is why such electrons are called "hot" electrons. In the B-B' diagram, some electrons with such high kinetic energy may be able to jump over the tunnel oxide barrier,

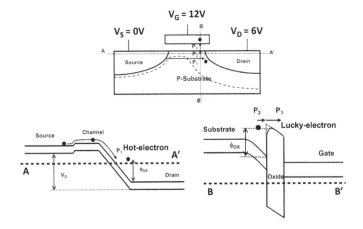

FIGURE 4.9 Illustration of the channel hot electron (CHE) effect, where the energy band diagrams along the channel direction (A-A') and perpendicular to the channel direction (B-B') are shown.

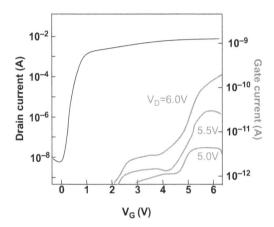

FIGURE 4.10 The I_D–V_G and I_G–V_G curves of a transistor at different V_D. The CHE effect introduces gate leakage at high drain voltages.

becoming "lucky" electrons. These electrons that are injected into the floating gate are considered lucky because the probability of thermally jumping over the barrier is very low. Figure 4.10 shows the I_D–V_G transfer curve of a transistor at different drain voltages and the gate current (I_G) is also plotted. Normally I_G is negligible at $V_G = 6$ V but as V_D is increased to 6 V, the CHE effect makes the gate current reach 50 pA. At the same time, the drain current is flowing at 5 mA. Therefore, the probability of becoming lucky electrons is given by the ratio of I_G over I_D. In this case, it is 50 pA/5 mA = 10^{-8}. Typically, the programming efficiency is very low 10^{-8}–10^{-6}. This is because when a large drain current flows (burning the power), only a very small fraction of electrons is injected into the floating gate.

4.2.4 Source-side Injection for Embedded Flash

CHE is generally used as the program mechanism for EEPROM and NOR Flash, which are primary types of embedded Flash (eFlash). The eFlash is compatible with the logic process and is widely used for code storage in microcontroller (MCU) products (e.g., for automotive electronics). The mainstream market for eFlash is at 40 nm node and above, and it is gradually moving toward the 28 nm platform; however, ultra-high-voltage transistors do not scale well with the logic process, and the fabrication cost (for extra masks) make eFlash scaling beyond 28 nm node remain elusive.

As aforementioned, CHE programming efficiency is very low. The programming efficiency is essentially determined by how many hot electrons could be generated and how many of them could be collected by the floating gate. However, there is a conflict between these two processes. A low gate voltage and a high drain voltage are preferred for a high electron generation rate as the lateral electric field drives the impact ionization. On the other hand, a high gate voltage and a low drain voltage are preferred for a high electron collection rate as the vertical electric field facilitates the thermal jump over the tunnel oxide barrier.

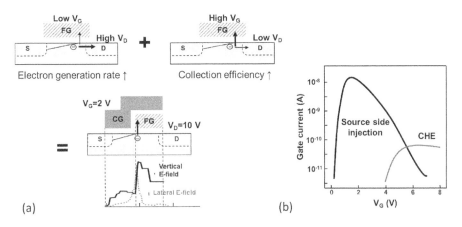

FIGURE 4.11 (a) The 1.5T split-gate structure with source-side injection (SSI) mechanism. (b) The gate current of the 1.5T split-gate transistor under SSI and its comparison with regular 1T floating-gate transistor under channel hot electron (CHE) effect.

One way to improve the programming efficiency is to use the source-side injection (SSI) method[3] with a one and a half transistor (1.5T) split-gate structure, as shown in Figure 4.11(a). The control gate that steps over the floating gate occupies approximately half of the gate length; therefore, this structure is named as 1.5T split-gate eFlash. A low voltage is applied on the control gate (e.g., 2 V), while a large drain voltage is applied (10 V), which effectively raises the floating gate to a high voltage due to the capacitive coupling (α_D will need to be designed close to 1 in this case by a large overlap region). Therefore, the electrons are laterally accelerated under the control gate, and a large volume of hot electrons is generated near the edge of the floating gate. These hot electrons now will see a large vertical electric field to facilitate jumping over the tunnel oxide barrier. Figure 4.11(b) shows the gate current of the 1.5T split-gate transistor under SSI and its comparison with the regular 1T floating-gate transistor under CHE. With the SSI method, the 1.5T split-gate could improve the programming efficiency to ~10^{-3}; thus, much lower power consumption is achieved. For the 1.5T split-gate eFlash, the erase could be done by F-N tunneling from the floating gate to the control gate (e.g., with large $V_{CG} = 15$ V) because the electric field is enhanced at the corner of the floating gate where it is stepped over by the control gate.

4.3 FLASH ARRAY ARCHITECTURES

4.3.1 NOR ARRAY

NOR is one of the commonly used Flash array architectures. The NOR Flash chip consists of many banks. For example, a 64 MB bank is divided into 1024 blocks, and each block is 64 kB. Then, one block consists of 16 sectors and each sector is 4 kB. Figure 4.12 shows the diagram for one sector of the NOR Flash, where the word lines (WLs) connect the control gates horizontally, and the bit lines (BLs) connect the drain of the floating-gate transistors vertically, and the source of the floating-gate

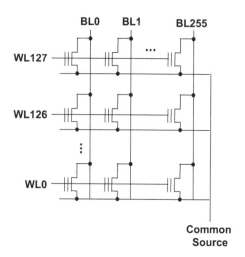

FIGURE 4.12 Diagram for one sector of the NOR Flash.

transistors are connected together by the common source. Typically, one WL is called one page. In this example, there are 128 pages, and each page has 32 Byte (i.e., 256 BLs) if one floating-gate transistor stores 1 bit. Essentially, the floating-gate transistors in one sector are all in parallel with each other. This is equivalent to the pull-down network of the NOR gates in the CMOS logic; thus, this type of array is referred to as the NOR Flash. The NOR Flash is usually programmable down to the individual bit. However, like all Flash memories, NOR is erased in large "chunks". Usually, the block is the largest group of bytes that can be erased (or flashed) in one operation, and the sector is the smallest group of bytes that can be erased (or flashed) in one operation.

Figure 4.13 shows the cross-section view of a BL of NOR Flash and the top view of a NOR Flash cell layout. To make a compact layout, usually, two adjacent floating-gate transistors share one BL contact via. The source lines are connected together on the silicon substrate as the sources are heavily doped. The top view of the layout shows a 10 F^2 cell area. The BL pitch is as minimum as 2 F (1 F for the wire, and 1 F for the isolation). The contact poly pitch (CPP) is designed as 5 F (1/2 F for each of the two contact vias, 1 F for each of the spacer, and 2 F for the gate length). Here, the gate length is not using the minimum length in order to sustain the large drain voltage, as the CHE mechanism is used for the NOR Flash program.

Figure 4.14 shows the NOR Flash's bias schemes for (a) erase, (b) program, and (c) read operations. The erase is done either with the source F-N tunneling or the channel F-N tunneling mechanism. One sector (or more sectors) that shares the same source line (or the same substrate) could be erased at the same time. For example, a high voltage of 12 V is applied to the common source while all the control gates are grounded. The electrons are ejected out from the floating gate with the source F-N tunneling for the erase. Typical erase time could be from 100 μs to 1 ms. The program is done with the CHE mechanism and could be performed at the individual bit level. For example, a high voltage of 12 V is applied to the control gate (the selected WL)

Flash Memory

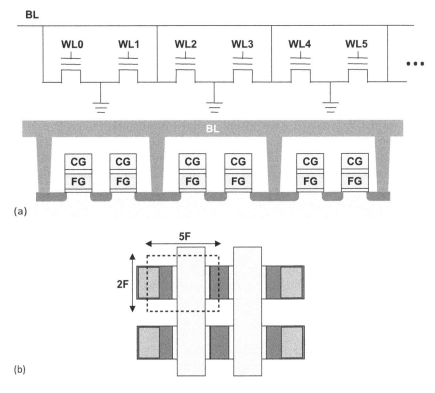

FIGURE 4.13 (a) The cross-section view of a BL of NOR Flash and (b) the top view of NOR Flash cell layout.

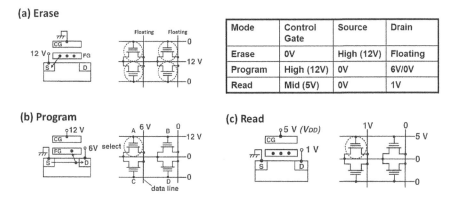

FIGURE 4.14 Representative NOR Flash's bias schemes for (a) erase, (b) program, and (c) read operations.

and a relatively large voltage of 6 V is applied to the drain (the selected BL) while the source is grounded. While a large drain current (~hundreds of μA) flows and a small fraction of hot electrons could inject into the floating gate for programming, in this case, only the selected cell A will be programmed. Cell B and Cell C are half-selected with either only the gate or the drain biased with high voltage. Ideally, they will not be programmed, but they are susceptible to the program disturb (i.e., slight change of the V_T value). This is because electrons may tunnel from the channel to the floating gate under high WL voltage in Cell B and electrons may tunnel from the floating gate to the drain under a relatively large drain voltage in Cell C. Cell D is unselected with zero voltages on all the terminals. Typical program time could be ~tens of μs. The read is done at the individual bit (or Byte) level. For example, a medium voltage 5 V is applied to the control gate (the selected WL) and a small voltage of 1 V is applied to the drain (the selected BL) while the source is grounded. If Cell A stores "0" in a programmed state (e.g., $V_{T\_H} = 5$ V), such 3 V read voltage could not turn on the channel; thus, the drain current is negligible. If Cell A stores "1" in an erased state (e.g., $V_{T\_L} = 1$ V), such 3 V read voltage could turn on the channel; thus, the drain current is detectable (i.e., discharging BL voltage and being sensed by sense amplifier). Typical read time could be reasonably fast down to ~50 ns.

4.3.2 NAND ARRAY

NAND is one of the commonly used Flash array architectures.[4] Figure 4.15 shows the 2D NAND Flash top-down organization for one plane. One NAND Flash chip typically consists of 1–4 planes, which can be independently operated. One plane generally consists of many blocks. For example, a 16-Gb single-level-cell (SLC) plane is divided into 2048 blocks and each block is 1 MB. Then, one block typically consists of dozens of pages. In this example, one block is divided into 64 pages (i.e.,

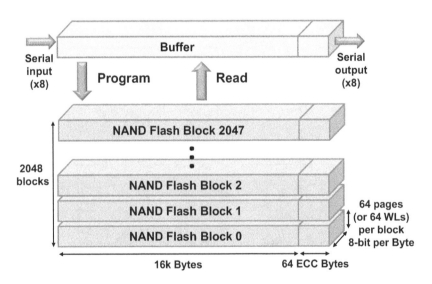

FIGURE 4.15 Example of organization chart of a 16-Gb plane for 2D SLC NAND Flash.

Flash Memory

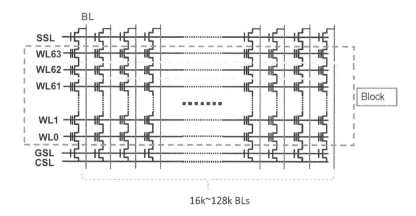

FIGURE 4.16 Diagram of a block for 2D NAND Flash.

64 WLs) and each page is 16 kB (i.e., 128k BLs) with additional parity bits for the error correction code (ECC). The typical page size could range from 2 kB to 16 kB. Figure 4.16 shows the diagram for one block where the WLs connect the control gates horizontally, the BLs connect one end of a string with floating-gate transistors in series, and the common source line (CSL) connects all the other ends of the strings to one block. The string has two selection gate transistors: the string select line (SSL) on top of the string and the ground select line (GSL) at the bottom of the string and they are regular MOSFETs (not floating-gate transistors). Essentially, the floating-gate transistors in one string are all in series. This is equivalent to the pull-down network of the NAND gates in the CMOS logic; thus, this type of array is referred to as the NAND Flash. Similarly, like all Flash memories, NAND must be erased in large "chunks". Unusually, one block or multiple blocks can be erased (or flashed) in one operation.

Figure 4.17 shows the cross-section view of a BL or string of 2D NAND Flash and the top view of a 2D NAND Flash cell layout. To make an ultra-compact layout, the floating-gate transistors in one string share the drain region with the adjacent neighbors, and the BL and the SL contact vias only exist at the two ends of the string. The top view of the layout shows a 4 F^2 cell area. The BL pitch is as minimum as 2 F (1 F for the wire, and 1 F for the isolation). The WL pitch is also minimum as 2 F because there is no contact via in between two neighboring cells. Therefore, the NAND Flash array achieves the theoretically highest density in a 2D plane.

Figure 4.18 shows the NAND Flash's bias schemes for erase operation to "1". The erase is done with the channel F-N tunneling mechanism. One block (or multiple blocks) that shares the same substrate could be erased at the same time. For example, a high voltage of 20 V is applied to the p-well substrate while all the control gates are grounded. The electrons are ejected out from the floating gate due to channel F-N tunneling. To avoid the breakdown of the string/ground select transistors, their gates are floating. The typical erase time of the entire block could be ~1 ms.

Figure 4.19 shows the NAND Flash's bias schemes for program operation to "0". The program is done with the channel F-N tunneling mechanism and is performed

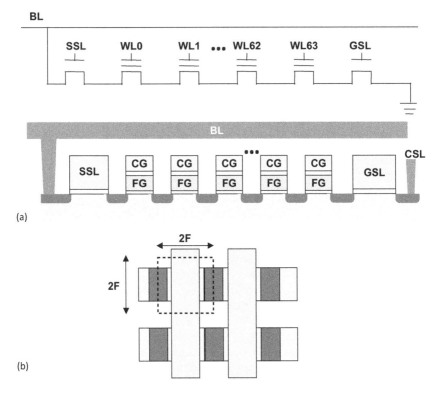

FIGURE 4.17 (a) The cross-section view of a BL or string of 2D NAND Flash and (b) the top view of 2D NAND Flash cell layout.

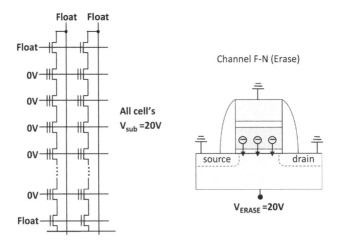

FIGURE 4.18 Representative 2D NAND Flash's bias scheme for erase of the entire block.

Flash Memory

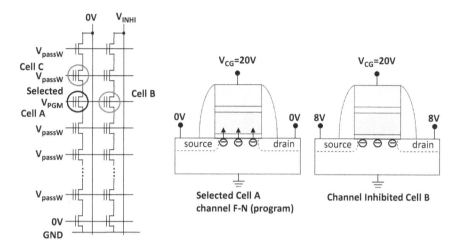

FIGURE 4.19 Representative 2D NAND Flash's bias scheme for program of the selected cell, and channel inhibition of the half-selected cell.

page by page. First, the gate of the SSL string select transistor is biased (e.g., at 10 V) and the gate of the GSL string select transistor is grounded (e.g., at 0 V). Second, a high program voltage (e.g., $V_{PGM} = 20$ V) is applied to the control gate of the selected WL (i.e., the selected page). In this example, Cell A in the selected page is selected and will be programmed to "0", and thus its associated BL should be grounded. On the other hand, Cell B in the selected page is half-selected and should remain at "1". To avoid the accidental write to Cell B, a "channel inhibition" scheme raises up its associated channel potential to some medium voltage, for example, 8 V. As a result, that the voltage drop across the control gate to the channel is reduced to 12 V in this case, which is much lower than the required $V_{PGM} = 20$ V. In order to pass the $V_{INHI} = 9$ V from BL to the Cell B, a pass voltage (V_{passW}) should be applied to other unselected WLs (or pages). Note that the string has a floating end (cut off by the GSL transistor); thus, the voltage passing along the string will follow the voltage clamping rule, $V_S = \min(V_G - V_T, V_D)$, which states that the source voltage potential (V_S) of a transistor is determined by the smaller one between the gate overdrive ($V_G - V_T$) and the drain voltage potential V_D. In other words, the $V_G - V_T$ should be larger than V_D for V_D to be effectively delivered to Cell B. If $V_D = V_{INHI} = 8$ V, and the programmed states V_{T_H} is 2 V, $V_{passW} = V_G$ should be at least $V_D + V_T = 10$ V. This suggests the unselected pages have V_{passW} applied on their control gates. For example, Cell C has 10 V across its control gate and channel. The typical program time for the selected page could be 10 µs–100 µs.

The conventional "channel inhibition" scheme as described above is not widely used anymore due to its high-power consumption as the unselected BL will be charged to a relatively large voltage, for example, $V_{INHI} = 9$ V. Alternatively, the "channel self-boost" scheme has been proposed. Figure 4.20(a)–(b) shows the timing diagram for the channel self-boost scheme and Figure 4.20(c) shows a simplified capacitive model. By charging the SSL and unselected BL to the voltage of the common collector V_{cc} (e.g., 3 V) first, then raising the V_{passW} to 10 V all the WLs, the string select transistors are closed since the channel potential is higher and makes the

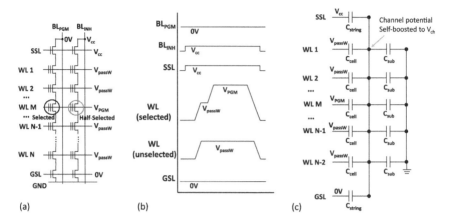

FIGURE 4.20 The channel self-boost scheme (a) for program inhibition and (b) associated timing waveform. (c) The simplified capacitive model for channel self-boost.

V_{GS} of the string select transistor essentially zero (V_G is V_{cc} and V_S is V_{cc}). At this point, the channel of the NAND string becomes a floating node. Next, by raising the selected WL to V_{PGM} = 20 V, the channel potential is boosted by the coupled series capacitance through the control gate, the floating gate, the channel, and the substrate. Using a simplified capacitive model (if neglecting the contribution from the top- and bottom-select transistors), the channel potential (V_{ch}) boost is given by

$$V_{ch} = \frac{C_{cell}}{N(C_{cell} + C_{sub})}(N-1)V_{passW} + \frac{C_{cell}}{N(C_{cell} + C_{sub})}V_{PGM} \quad (4.12)$$

where C_{cell} is the combined series capacitance of the C_k for the inter-poly dielectric capacitance and C_g for the tunnel oxide capacitance, and C_{sub} is the depletion capacitance of the channel to the substrate. N is the number of WLs in one NAND string. This raised V_{ch} makes the effective voltage drop on the half-selected cell ($V_{PGM}-V_{ch}$) not being programed.

To further increase the channel self-boost efficiency, the local self-boost scheme and the asymmetric self-boost scheme are developed, as shown in Figures 4.21(a) and (b), respectively. The local self-boost scheme biases the adjacent WLs (upper and lower) of the selected WL to ground after the channel becomes a floating node. Hence, the selected cell is isolated from other cells in the string; thus, the boosted channel potential of the half-selected cell could be higher because the charges no longer need to be shared with other cells. The channel potential (V_{ch}) boost is given by

$$V_{ch} = \frac{C_{cell}}{(C_{cell} + C_{sub})} V_{PGM} \quad (4.13)$$

One problem of the local self-boost scheme is the grounded upper WL makes it difficult to pass the zero voltage to the selected BL for the programming of the

Flash Memory

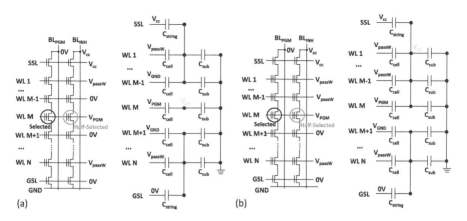

FIGURE 4.21 (a) The local self-boost scheme and (b) the asymmetric self-boost scheme.

selected cell, which may require a more complex timing design for selected BL and unselected BL separately. A trade-off design is to only isolate the lower section of the NAND string, namely the asymmetric self-boost scheme, where only the adjacent WL that is lower to the selected WL is grounded. Hence, the selected cell is isolated from other cells in the upper section of the string. If there are M cells in the upper section of the string (M<N), the channel potential (V_{ch}) boost is given by

$$V_{ch} = \frac{C_{cell}}{M(C_{cell} + C_{sub})}(M-1)V_{passW} + \frac{C_{cell}}{M(C_{cell} + C_{sub})}V_{PGM} \quad (4.14)$$

Apparently, the boost efficiency of the asymmetric self-boost scheme is in between the self-boost scheme and the local self-boost scheme.

Figure 4.22 shows the NAND Flash's bias schemes for the read operation. The read is done page by page. First, the gate of the SSL is biased (e.g., 4 V) and the gate of the GSL string select transistor is also biased (e.g., at 4 V) to allow a possible current path from BL to CSL where BL is pre-charged (e.g., $V_{BL} = 0.3$ V). Second, for the selected page, a read voltage V_R (e.g., 0 V) that is in between the programed V_{T_H} (e.g., 2 V) and the erased V_{T_L} (e.g., −2 V) is applied to the selected WL while the source is grounded. It is important to turn on other unselected pages to "pass" the current regardless of their memory states. Therefore, a pass voltage (V_{passR}, e.g., 5 V) that is higher than the V_{T_H} is applied to all the other unselected WLs. If the selected Cell A stores "0" in a programmed state (e.g., $V_{T\_H} = 2$ V), the read voltage ($V_R = 0$ V) could not turn on the channel; thus, the entire string will conduct negligible current (as Cell A is blocking other transistors in series) and V_{BL} will not decay. If the selected Cell A stores "1" in an erased state (e.g., $V_{T\_L} = -2$ V), the read voltage ($V_R = 0$ V) could turn on the channel; thus, all the transistors along the entire string are conducting (as there is a current flowing from BL to CSL along the string) and V_{BL} will decay. Because the series resistance of the discharge path is a sum of the channel resistance of all the transistors (e.g., 64) along the string, the parasitic RC delay is substantial. Therefore, the typical read time could be as long as 10–100 μs.

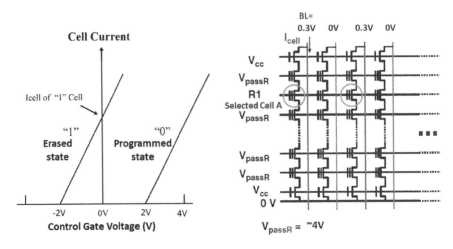

FIGURE 4.22 Representative 2D NAND Flash's bias scheme for the read of the selected BLs.

4.3.3 Peripheral Circuits for High Voltage

Since Flash uses high voltages (e.g., 20 V or above) for program/erase, the peripheral circuits need to support such high-voltage operations. The basic building blocks of a high-voltage system are the charge pump circuit and the level shifter circuit. The charge pump circuit is able to generate a high-voltage DC bias using an external low-voltage power supply. The level shifter is able to switch the input signals from the low-voltage domain to the high-voltage domain as the output signals. Figure 4.23(a) shows a simple design for the voltage doubler. By applying the VDD to input and VDD/GND clock pulse to CK and CK# complementary, MN1 or MN2 is turned on then V_A and V_B are charged and discharged between VDD and 2 × VDD. According to the V_A and V_B voltage, MP1 or MP2 pass the 2 × VDD to the output. The output voltage is doubled as the input voltage. In principle, such voltage doubler can be cascaded to several stages to further boost the voltage to the desired high-voltage level. More advanced techniques are needed to stabilize the output voltage (e.g., to reduce the ripple effect) to make it an ideal DC voltage bias source. Figure 4.23(b) shows

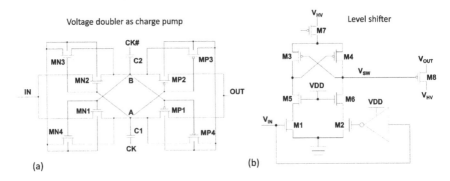

FIGURE 4.23 (a) Circuit schematic of the voltage doubler as the charge pump. (b) Circuit schematic of the level shifter.

Flash Memory

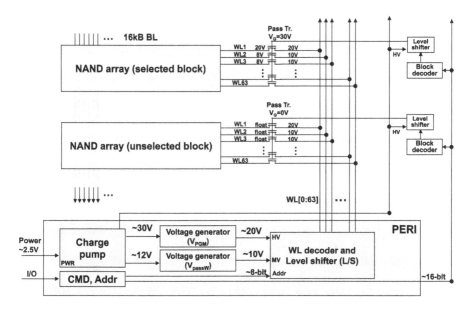

FIGURE 4.24 The high-voltage peripheral system for NAND Flash.

a simple design for the level shifter. If V_{IN} is grounded, M2 is turned on and V_{SW} is grounded; therefore, the V_{OUT} is equal to V_{HV}. On the other hand, if V_{IN} is V_{DD}, M4 is turned on and V_{SW} is equal to V_{HV} which means M8 is turned off; therefore, V_{OUT} is equal to the ground The high-voltage bias is from the charge pump circuit. The input signal is toggling between the low-voltage power supply and the ground, while the output signal is toggling between the high-voltage power supply and the ground.

Figure 4.24 shows the high-voltage peripheral system for NAND Flash, illustrating where the charge pump and the level shifter are used for the program operation. First, the external power supply voltage (e.g., 2.5 V) is converted to V_{PGM} (e.g., 20 V) and V_{passW} (e.g., 10 V), and they are sent to a level shifter that is after the WL decoder as voltage supplies. According to the WL address bits, only one WL is biased with V_{PGM}, and the other WLs are biased with V_{passW} (to pass the BL voltage). Then, the WL voltages are sent to all the blocks using a data bus (e.g., with 64-channels). On the other hand, the block address bits select the block through another block decoder, and only the block that is being programmed is boosted to a high voltage (e.g., 30 V) after the level shifter. This high voltage is able to turn on the pass-transistors that control the WLs in the selected block and V_{PGM} and V_{passW} are delivered through. Other blocks are not selected with 0 V on the pass-transistors.

4.3.4 NAND Flash Translation Layer

Due to the unique features of page-based program/read and block-based erase, the NAND Flash is not a random access memory. The program is sequentially performed from one page to the next page within a block, while the erase is done for the entire block. This means that the entire block should start with all cells in the erased states (all "1"s), then program operation will write a page with certain strings to "0"s and remain the rest at "1"s depending on the data pattern that is to be stored. A write pointer is used to label the current page that is being selected. The write pointer will move sequentially page by page in a selected block (i.e., from page 0 to page 63) and this order cannot be reversed, or the pointer cannot jump from one block to another block if it does not reach the last page (i.e., page 63). Hence, a mapping relationship exists from the logic block address (LBA) that is seen by the host to the actual physical page address in NAND Flash. In the NAND controller, there is a Flash translation layer (FTL) that serves as a logbook to record such mapping and additional information about the physical blocks. Figure 4.25 shows an example of an FTL information

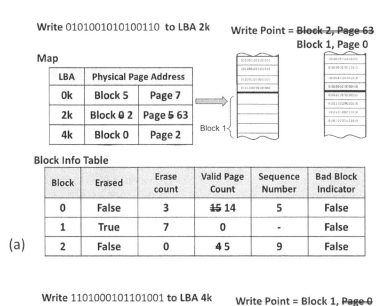

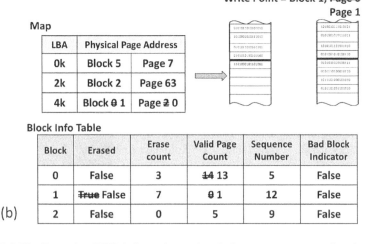

FIGURE 4.25 Example of FTL information update during a program operation that jumps from (a) the end of one block to (b) the beginning of another block.

update during a program operation that jumps from one block to another (assuming page size is 2 kB). Figure 4.25(a) shows the current write pointer is at block 2 and page 63. Now the controller needs to write a new data pattern to the LBA starting from 2k in NAND Flash, where this LBA was previously stored at physical block 0 and page 5. Now the new data pattern for LBA 2k will be programmed to physical block 2 and page 63 where the pointer is currently located. The map will be updated to record this mapping. Then, the write pointer will need to move to a new "empty" block (e.g., physical block 1)'s page 0. Meanwhile, the block info table in FTL will update such information. For example, the valid page count in physical block 0 will reduce by 1 while in physical block 2 will increase by 1. Figure 4.25(b) shows the next step when the new data pattern for LBA 4k will be programmed to physical block 1 and page 0 where the pointer is currently located. The map will be updated to record this mapping. Meanwhile, the block info table in FTL will update such information. For example, the valid page count in physical block 0 will reduce by 1 while in physical block 0 will increase by 1. Since this is a new physical block that is being programmed, the indicator for "erased" will be labeled as false. Then, the write pointer will need to move to physical block 1 and page 1.

Since the program operations always consume new physical blocks. Over time, page invalidations cause fragmentation within a block, where most pages in the block are invalid. The FTL periodically performs garbage collection, which identifies each of the highly fragmented Flash blocks and erases the entire block (after migrating any remaining valid pages to a new block). Hence, at a certain point, the controller needs to schedule erase operations to a block or multiple blocks to ensure there are enough erased "empty" blocks to be ready for programming (in case a large chunk of data needs to be programmed in the real workloads). Figure 4.26 shows an example of an FTL information update during an erase operation (e.g., targeting at physical block 2). Figure 4.26(a) shows the current write pointer is at block 1 and page 4. Before physical block 2 is erased, all the valid pages in this physical block should be transferred out. Therefore, these valid pages in physical block 2 should be read out and then programmed to the other block where the pointer is currently located (in this case, block 1 page 4 to page 6). Meanwhile, the block info table in FTL will update such information. For example, the valid page count in physical block 2 will reduce to 0 while in physical block 1 will increase by 3. Then, the entire block 2 is erased to all "1"s, as shown in Figure 4.26(b). The block info table will update relevant information of block 2 such as the erase count increases by 1 and the erased indicator is set to be true.

The FTL keeps track of the mapping and information of all the physical blocks. The LBA, that is continuous conceptually, may be stored in different pages or different blocks physically. The NAND Flash program and erase is not as flexible as the random access memory, creating much more complexity in the controller design.

4.3.5 COMPARISON BETWEEN NOR AND NAND

Table 4.1 summarizes the key differences between NOR Flash and 2D NAND Flash. 2D NAND has a smaller cell size ($4\ F^2$) than that of the NOR ($10\ F^2$). 2D NAND has been scaled to around 15 nm feature size, while NOR stays at a feature size of 55 nm

Block Info Table

Block	Erased	Erase count	Valid Page Count	Sequence Number	Bad Block Indicator
0	False	3	13	5	False
1	False	7	1	12	False
2	False	0	~~3~~ 0	9	False

(a)

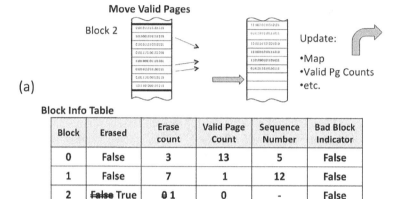

Block Info Table

Block	Erased	Erase count	Valid Page Count	Sequence Number	Bad Block Indicator
0	False	3	13	5	False
1	False	7	1	12	False
2	~~False~~ True	~~0~~ 1	0	-	False

(b)

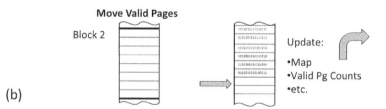

FIGURE 4.26 Example of FTL information update (a) before and (b) after an erase operation.

TABLE 4.1
Summary of the Key Differences between NOR Flash and 2D NAND Flash

SPECs	2D NAND	NOR
Density	4 F^2 (SLC)/1.3 F^2 (TLC)	10 F^2
Technology node	14 nm ½ pitch (2015)	55 nm ½ pitch (2015)
Scaling issues	Fewer electrons/capacitance crosstalk	Fewer electrons/Short channel effect
Cost per bit	Low	High
Data Access mode	Serial access	Random access
Program/Erase	100 µs–1 ms/page	10 µs/Byte
Read	25 µs/page	50 ns, random access
Endurance	10^3~10^4 cycle	>10^5 cycle
Retention	2–5 years	10 years
Application	Mass storage	Code storage

or above in 2015. In addition, NAND Flash is more aggressive to employ multilevel cell (MLC), for example, 3 bits per cell or even 4 bits per cell. As a result, NAND has a much higher density (and lower cost per bit) than NOR (even before NAND switches to 3D integration). 2D NAND could offer up to 128 Gb per chip and 3D NAND could offer 1 Tb per chip as of 2020, while standalone NOR Flash generally offers 256 Mb to 8 Gb per chip.

As discussed earlier, NAND has to reply on the sequential access, but NOR can enable random access. The random access read speed of NOR is relatively short (~50 ns) as each floating-gate transistor is directly attached to the BL. In contrast, the random access read speed of NAND is relatively long (~10 μs). However, when the first page of the NAND is read out, the rest of the pages in the selected block are relatively fast (~50 ns) because the I/O could burst out the pre-fetched data from a large number of BLs in one page. Typically, the reliability of NOR Flash has a higher standard than that of NAND Flash. For example, NOR Flash can sustain 10^6 program/erase endurance cycles and more than 10 years of data retention at elevated temperature (e.g., 85 °C), while NAND Flash can only sustain 10^3–10^4 cycles and ~3 years data retention. Despite these advantages of NOR Flash, its relatively higher cost makes its market being taken away by NAND Flash. Today NAND Flash dominates in the massive data storage (e.g., images, videos, files, etc.). The major vendors of NAND Flash as of 2020 include Samsung, SK Hynix (which will acquire Intel's NAND business), Micron, Kioxia, and Western Digital.

The market of NOR Flash is about $2 billion in 2020 (about 3.5% of the NAND Flash). The primary vendors for standalone NOR Flash include Macronix, Winbond, Cypress, Micron, and GigaDevice. In general, NOR Flash offers lower capacity, fast random read access, and relatively higher data reliability as is required for code storage and execution. NOR Flash allows SoC (system-on-chip) and FPGA (field-programmable gate array) to boot from it directly; thus, it is commonly used to store the firmware. Today's standalone NOR Flash is mostly offered in 40 nm, and the scaling is almost stopped, urging the development of emerging non-volatile memories (to be discussed in Chapter 5).

4.4 MULTILEVEL CELL

4.4.1 Multi-level Cell (MLC) Basics

Flash memory (especially for NAND) aims for improving integration density by storing multiple bits of data into one floating-gate transistor or one cell. Specifically speaking, 2 bits per cell is referred to as the multi-level cell (MLC),[5] 3 bits per cell is referred to as the triple-level cell (TLC), and 4 bits per cell is referred to as the quadruple-level cell (QLC). The conventional Flash is operated in single-level cell (SLC) mode with a programmed V_T state and an erased V_T state. The way to realize MLC is to modulate the threshold voltage into multiple levels in the partially programmed intermediate states. Figure 4.27 shows the conceptual V_T distribution of the Flash memory cells in arrays for SLC, MLC, and TLC. Generally, for NAND Flash, the erased V_T is negative (e.g., around −2 V), while the fully programmed V_T is positive (e.g., around +4 V). There is a big gap between the two V_T distributions where

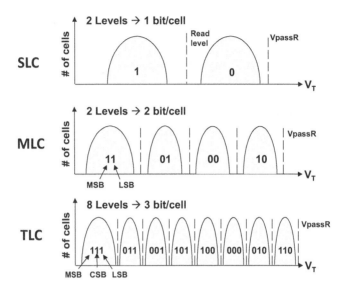

FIGURE 4.27 The V_T distribution of the Flash memory cells in arrays for SLC, MLC, and TLC.

intermediate states could be inserted to encode more bits. For MLC, four levels of V_T (ER, P1, P2, and P3, from low V_T to high V_T) are required to encode 11, 01, 00, and 10, respectively. For TLC, 8 levels of V_T (ER, P1, P2, P3, P4, P5, P6, and P7, from low V_T to high V_T) are required to encode 111, 011, 001, 101, 100, 000, 010, and 110, respectively. Note that some manufacturers may choose to use a different encoding of V_T levels to the binary bits. In principle, the bit values of adjacent levels are separated by a Hamming distance of 1. This is helpful to minimize the error rate because if there is any shift of the V_T from one level to its adjacent level, it will only result in a 1-bit error. Since the memory window between the erased state and the fully programmed state is constrained (e.g., around 6 V). The distance between V_T levels is much reduced in TLC than in MLC. Considering the spread-out of V_T distribution caused by process variations or programming inaccuracy, the distance between adjacent levels could be as small as 600 mV for TLC. Recently, the industry has moved toward QLC with 4 bits per cell, requiring 16 levels of V_T distribution that are tightened, and the distance between adjacent levels could be as small as 300 mV for QLC.

The method to program the Flash memory to the intermediate V_T level is to adjust the programming voltage amplitude and the programming pulse width. The underlying physics is to modulate the number of electrons that are injected into the floating gate. As aforementioned, the more electrons in the floating gate, the higher V_T is. If the number of electrons being injected could be precisely controlled, then multiple V_T levels could be realized. Figure 4.28 shows an example of the V_T as a function of program time (1 μs to 1 ms) under different program voltages (17 V to 19 V). The V_T monotonically increases with either the program time or the program voltage. Therefore, the program time and the program voltage are the primary knobs in tuning the V_T for MLC, TLC, and QLC.

Flash Memory

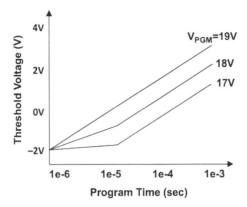

FIGURE 4.28 Example of the V_T as a function of program time under different program voltages for a Flash transistor.

4.4.2 Incremental Step Pulse Programming (ISPP)

Because there exists process variation and programming inaccuracy, it is almost impossible to achieve a precise V_T level after a single programming pulse. Hence, a write-and-verify scheme is necessary to achieve the targeted V_T level for MLC. TLC and QLC are more demanding for a tight V_T distribution as the margin between levels is much narrower. A write-and-verify scheme will apply one programming pulse and then read the memory state out and compare to the target V_T level. If the current state does not meet the criterion of the target V_T range, an additional pulse will be applied till meeting the criterion. To avoid over-programming (to larger than the target V_T), relatively low pulse amplitude should be applied first, but it may take many pulses to reach the target.

To address this issue, Incremental Step Pulse Programming (ISPP) is a commonly used write-and-verify scheme for MLC/TLC/QLC programming. The pulse amplitude is gradually increased over the pulsing sequence. Figure 4.29 shows an example of ISPP for programming toward an intermediate V_T level. Using an ISPP condition (initial amplitude 16 V with a step size of 0.5 V), 2–6 program/verify cycles are

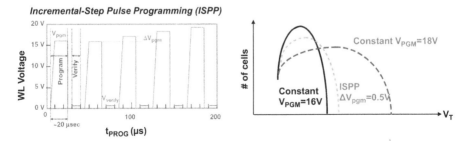

FIGURE 4.29 Example of ISPP for programming waveform and the resulting V_T distribution compared to a write-and-verify scheme with constant program voltages.

needed to tighten the V_T distribution with $3\sigma = 0.6$ V. In contrast, if a constant $V_{PGM} = 16$ V is used during write-and-verify, it may take 11–37 program/verify cycles to achieve a similar V_T distribution with $3\sigma = 0.5$ V; if constant $V_{PGM} = 18$ V is used during write-and-verify, the resulting distribution is much spread out with $3\sigma = 1.5$ V as over-programming dominate the tail. In sum, ISPP provides a fast and effective programming scheme to tighten the V_T distribution. It should be pointed out that ISPP can only inject more electrons into the floating gate but cannot partially remove electrons from the floating gate; thus, it is impossible to modify data to a new arbitrary value or switch the data arbitrarily between two intermediate states. Hence, a complete erase is necessary before a new data pattern could be written into the Flash.

The practical programming protocol from the erased states to the intermediate programmed states may take a staged programming sequence. Figure 4.30(a) illustrates the two-step programming protocol for MLC. In the first step, a Flash transistor is partially programmed based on its least significant bit (LSB) value, either staying in the ER state if the LSB value is "1" or moving to a temporary state (TP) if the LSB value is "0". The TP state has a mean voltage that falls between levels P1 and P2. In the second step, the LSB data is first read back into an internal buffer register to determine the cell's current threshold voltage state, and then further programming pulses are applied based on the most significant bit (MSB) data to increase the cell's V_T to fall within the voltage window of its final state. ER becomes either P1 or remains at ER, and TP becomes either P2 or P3.

In principle, TLC takes a similar approach to the two-step programming of MLC. One of the commonly used schemes is known as foggy-fine programming, which is illustrated in Figure 4.30(b). The Flash transistor is first partially programmed based on its LSB value, using the binary programming step in which the coarse-grain ISPP step-pulses are used to significantly increase the V_T level to the TP state. Then, the Flash transistor is partially programmed again based on its central significance bit (CSB) and MSB values to a new set of temporary states (these steps are referred to as foggy programming, which uses the finer-grain ISPP step-pulses than the

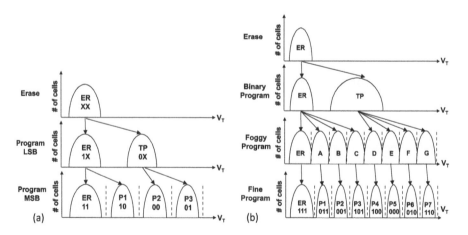

FIGURE 4.30 Programming sequence for (a) MLC and (b) QLC.

binary programming). Due to the higher potential for errors during TLC programming caused by the narrower margin, all of the programmed bit values are buffered after the binary and foggy programming steps into the internal buffer register. Finally, fine programming takes place, where these bit values are read from the buffers, and the smallest ISPP step-pulses are applied to set each cell to its final threshold voltage level. The purpose of this last fine programming step is to strictly tighten the V_T distributions.

Obviously, complex programing schemes take time. For example, MLC (~200 μs/page) and TLC (~500 μs/page), and QLC (up to 2 ms/page) are more time consuming than SLC (less than 100 μs/page). Hence, there is a trade-off between the integration density and the speed.

4.5 FLASH RELIABILITY

Flash memory may suffer from several reliability-degradation mechanisms: 1) endurance degradation after program/erase (P/E) cycling; 2) retention after a long time (especially at elevated temperature); 3) disturb after program/read operations. As a result, the V_T distribution may deviate from its original shape or position, leading toward shifted and widened distribution compared to their initial states. Due to the shift and widening, the tails of the two neighboring V_T distributions may overlap with each other. The read reference voltages can no longer correctly identify the state of the Flash cells in the overlapping region, leading to errors during a read operation. There are two types of errors: 1) hard error and 2) soft error. A hard error means that the error is permanent. For example, the V_T value is stuck at a certain level and could not be further changed, which typically occurs at endurance failure at the end of the lifetime. A soft error means that the error is recoverable. For example, the V_T value shifts over time, but after an erase operation, it could be reprogrammed to its original value. This is usually the case for the retention degradation or the disturb degradation.

Generally, Flash memory is equipped with ECC to logically correct the raw errors resulting from these reliability degradation mechanisms. For example, a page of 2 kB has additional 64 Bytes at the end of the page as the parity bits. The tolerable raw bit error rate (BER) can be 10^{-4}–10^{-2}. After the ECC, the post-correction bit error rate should approach 10^{-15}. If the prolonged P/E cycling, excessive-high temperature over a long time, or continuous program/read disturb causes too many errors that are beyond the correction capability of the ECC, a true error will occur, which sets the limit of the lifetime of the Flash memory product.

4.5.1 Endurance

Endurance means how many times the memory cell can be programmed and erased before the memory window collapses. Endurance degradation is also called wear-out. As more write operations are performed to a block, there are a greater number of raw bit errors that must be corrected, which may exceed the correction capability of the ECC. Figure 4.31 shows a representative V_T evolution as a function of P/E cycling up to 10^6 cycles for the SLC mode. The erased state V_T keeps increasing over P/E cycling and the margin between the two states shrinks. The underlying physical

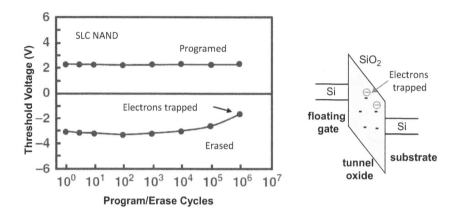

FIGURE 4.31 Example of V_T evolution as a function of P/E cycling up to 10^6 cycles for the SLC NAND and schematic of electrons trapped in the tunnel oxide.

mechanism is attributed to the excessive electron trapping in the tunnel oxide. As the P/E cycling progresses, electrons are injected into and ejected out of the floating gate many times. The tunneling process has a finite probability to damage the tunnel oxide, by creating defects (i.e., oxygen vacancies). The defects act as traps that can capture electrons. The number of defects grows over P/E cycling; thus, more and more electrons are captured into the tunnel oxide, which are not easy to be removed during the erase operation. As a result, incomplete removal of electrons in the tunnel oxide effectively increases V_T.

For MLC (or TLC/QLC), the endurance degradation causes more errors as the margin between adjacent levels is relatively small. Figure 4.32 shows the V_T distributions of MLC NAND Flash memory before and after P/E cycles. Two observations could be made. First, as the P/E cycle count increases, each state's V_T distribution systematically (1) shifts to the right and (2) becomes widened. Second, the amount of the shift is greater for lower V_T states than it is for higher V_T states. The erased state's V_T increases due to the incomplete removal of electrons induced by trapping. The programmed V_T distribution shift occurs because as more P/E cycles take place, the quality of the tunnel oxide degrades, allowing electrons to tunnel through the oxide more easily (e.g., via trap-assisted tunneling). As a result, if the same ISPP conditions (e.g., programming voltage, step-pulse size, program time) are applied

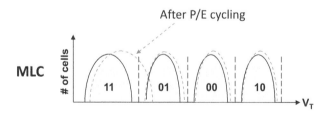

FIGURE 4.32 Illustration of the V_T distributions of MLC NAND before and after P/E endurance cycles.

Flash Memory

throughout the lifetime of the NAND Flash memory, more electrons are injected during programming as a Flash memory block wears out, leading to higher V_T, i.e., the right shift of the distribution. The distribution of each state widens due to the process variation present in the wear-out process. As each V_T distribution widens, more overlap occurs between neighboring distributions, making it less likely for a read reference voltage to determine the correct value of the cells in the overlapping regions.

The endurance specification varies significantly among NAND Flash products. The 50 nm node of SLC NAND could endure ~10^5 P/E cycles per block. The 50 nm node of MLC NAND flash could endure ~10^4 P/E cycles per block, the more advanced sub-20 nm generations of MLC and TLC NAND Flash can endure only ~3000 and ~1000 P/E cycles per block, respectively. Stronger ECC and smart wear-leveling techniques are thus needed for scaled and higher density NAND Flash.

4.5.2 RETENTION

Retention means how long the charges could be retained in the floating gate and the memory window can maintain between states. Retention errors are caused by charge leakage over time and reflected as the V_T distribution drifts over time. There are two types of charge leakage mechanisms. The first type is intrinsic charge loss, which is mainly through thermionic emission over the energy potential barrier of the tunnel oxide between the floating gate and the substrate. The second type is extrinsic charge loss, which includes two mechanisms: 1) charge de-trapping and 2) trap-assisted tunneling, as shown in Figure 4.33. Charge de-trapping, where charges previously trapped in the tunnel oxide after P/E cycling are freed spontaneously, can also occur over time. As the trapped charges are mostly electrons in the tunnel oxide, the V_T distribution tends to drift negatively over time as the electrons are being released to the substrate. Trap-assisted tunneling occurs because the defects that are generated over P/E cycles form step stones for enhanced tunneling probability, which exacerbates the stress-induced leakage current (SILC). As SILC is bidirectional, it can either assist electrons to tunnel out of or into the floating gate depending on the voltage polarity. Hence, V_T distribution may see a negative drift for the programmed states and a positive drift for the erased states. As the Flash memory wears out with increasing P/E cycles, the amount of trapped charges also increases, and so does the SILC effect. At prolonged P/E cycles, the amount of trapped charges is sufficiently large to form percolation paths that significantly hamper the insulating properties of the tunnel oxide, resulting in significant tail bits of the V_T distribution. Figure 4.34 shows the V_T distribution drift under the de-trapping and trap-assisted tunneling effects.

Both the intrinsic thermionic emission and the de-trapping process are thermally activated; thus, the retention can be characterized by a varying-temperature acceleration measurement. Figure 4.35 shows the typical retention baking testing results for an SLC NAND Flash transistor. Different temperatures of 200 °C, 250 °C, and 300 °C are applied, and the V_T of the programmed state (initially = 3 V) is monitored in Figure 4.35(a). If a pre-defined criterion of memory window is used (e.g., V_T = 1 V), the retention time-to-failure could be determined as the V_T drift by 2 V. Then, the retention time-to-failure in the logarithmic scale could be plotted with 1/kT,[6] namely the Arrhenius plot as shown in Figure 4.35(b). The slope is extracted as the activation

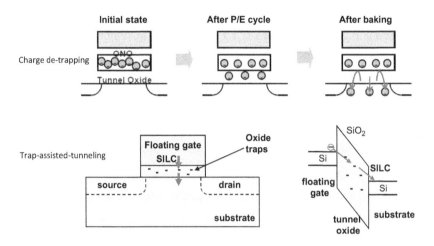

FIGURE 4.33 Illustration of extrinsic charge loss mechanism in floating-gate transistor showing the charge de-trapping and the trap-assisted tunneling.

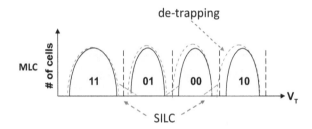

FIGURE 4.34 Illustration of the V_T distribution drift of MLC NAND under the de-trapping and trap-assisted tunneling effects during retention.

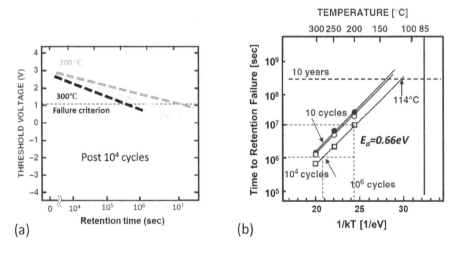

FIGURE 4.35 (a) Example of V_T drift of the programmed state of an SLC NAND under high-temperature baking. (b) The Arrhenius plot for time-to-retention failure.

energy (E_a) in the unit of electron volt (eV) and, in this case, $E_a = 0.66$ eV. The extrapolation is used to estimate the data retention lifetime whether exceeds 10 years at a certain baking temperature. Typically, post-P/E cycling, the retention degrades faster due to the enhanced de-trapping and SILC effects. In this example, the extrapolation line gives an estimated 10-year retention time at 114 °C for this Flash transistor if it already endures 10^4 P/E cycles. It should be noted that MLC (TLC/QLC) will have a much narrower margin between levels for V_T drift.

4.5.3 Disturb

Disturb means when the selected cell is being programmed or read, the disturbance on the neighboring cells' states occurs in terms of the V_T shift. Figure 4.36 shows the program disturb pattern of a NAND Flash array and the V_T shift for the half-selected cells. The cells that share the same BL with the selected cell will suffer from the program disturb due to the V_{passW} applied on unselected WLs. Because V_{passW} is a medium voltage (e.g., ~10 V), the F-N tunneling probability is significantly lower than the actual programming. However, there is still a finite probability that extra electrons tunnel into the floating gate, leading to an increase of V_T. On the other hand, the cells that share the same WL with the selected cell will suffer from the program disturb due to the voltage drop (e.g., $V_{PGM} - V_{ch} \sim 12$ V). Similarly, there is still a finite probability that extra electrons tunnel into the floating gate, leading to an increase of V_T. The conflict of the unselected WL and unselected BL leads to a constrained margin of the V_{passW}. If V_{passW} is too large (e.g., above 12 V), the unselected WL will be under severe program disturb and its V_T will increase substantially. If V_{passW} is too small (e.g., below 11 V), the channel inhibition or the channel self-boost is no longer effective and the unselected BL will be under severe program disturb and its V_T will increase. Hence, the margin for V_{passW} is from 11 V to 12 V in this example. Overall, the tail bits in the V_T distribution will shift toward positive under the repeated program disturb, as shown in Figure 4.37(a). The shift is more substantial for the low V_T

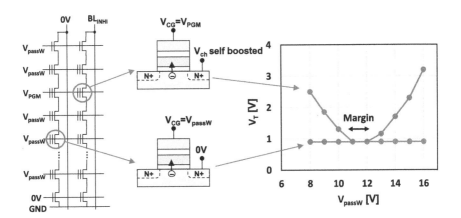

FIGURE 4.36 The program disturb pattern of a NAND Flash array and the V_T shift for the half-selected cells.

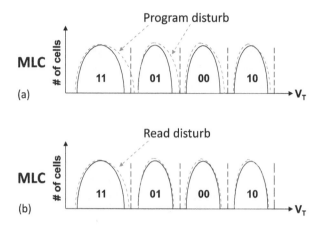

FIGURE 4.37 Illustration of the V_T distribution shift of MLC NAND under (a) the program disturb and (b) the read disturb.

states such as the erased state or P1 state because they see a relatively larger electric field across the tunnel oxide.

The read disturb may also occur in the NAND Flash array. The cells that share the same BL with the selected cell will suffer from the read disturb due to the V_{passR} applied on unselected WLs. Because V_{passR} is a relatively low voltage (e.g., ~6 V for MLC/TLC/QLC), the F-N tunneling probability is significantly lower than the actual programming. However, there is still a finite probability that extra electrons tunnel into the floating gate, especially when repeated read operations are performed. Hence, the tail bits in the V_T distribution will shift toward positive under the repeated read disturb, as shown in Figure 4.37(b). The shift is mostly substantial for the low V_T states (i.e., the erased state) because they see a relatively larger electric field across the tunnel oxide. Compared to the program disturb, the impact of the read disturb is less problematic.

4.5.4 Trade-offs between Reliability Effects

To summarize the reliability effects, it is noted that retention and program/read disturb interact with the endurance. Generally, retention failure and program/read disturb become severe after extensive P/E cycling, because of the excessive defect/trap generation in the tunnel oxide that facilitates the trap-assisted tunneling and de-trapping. As a result, trade-offs exist between the reliability specifications and multi-level operations. For example, the ideal specifications of SLC NAND are endurance = 10^5 cycles and retention = 10 years extrapolated at 85 °C. However, these two specifications cannot be met at the same time especially for MLC/TLC/QLC NAND. For example, the TLC specifications may reduce to endurance = 10^3 cycles and retention = 3 years extrapolated at 85 °C. The post-cycling may also negatively affect data retention. For the fresh chips with a few numbers of P/E cycling, the retention may exceed 10 years extrapolated at 85 °C. Toward the lifetime end of P/E cycling (e.g., after 10^3 cycles), the retention may degrade to 1 year at 85 °C.

4.6 FLASH SCALING CHALLENGES

Since its invention in the mid-1980s, 2D NAND Flash scaling has been progressed well and even outpaced the logic transistor's scaling, and it has become the driving force for advanced lithography technology in the late 2000s. The feature size has been scaled from ~1.5 μm in 1986 to ~15 nm in 2016. That is 100× in dimensional scaling and 10,000× in area scaling in 3 decades. Coupled with MLC/TLC, the NAND chip capacity also steadily increases from 1 Gb in 2001 to 128 Gb in 2016, as shown in Figure 4.38. That is a 128× density improvement in 15 years. Nevertheless, several grand challenges arise in 2D NAND Flash scaling in the mid-2010s besides the limitations from the lithographic patterning in fabrication. The sub-20 nm generations include three nodes, 1x nm, 1y nm, and 1z nm.[7] From this chart, it is also shown that the chip capacity stays at 126 Gb in these three nodes, though the chip area shrinks. The 1z node is the last generation of the 2D NAND Flash. Two primary challenges that limit the 2D NAND Flash's continued scaling are 1) the cell-to-cell interference and 2) the few electrons problem, as discussed in the following sections.

4.6.1 Cell-to-Cell Interference

In a NAND Flash array, the floating-gate transistor is capacitively coupled to its neighboring cells, as shown in Figure 4.39. Due to WL-to-WL, BL-to-BL, and diagonal coupling by parasitic capacitances, the selected cell in the middle electrostatically interferes with its neighbors during the programming. Hence, the V_T may be shifted by the neighboring cell's high V_{PGM}. With the scaling, the distance between cells becomes shorter, and thus the parasitic capacitances increase. In sub-20 nm 2D NAND generations, the V_T shift could be up to 40% if the cell's all the neighboring

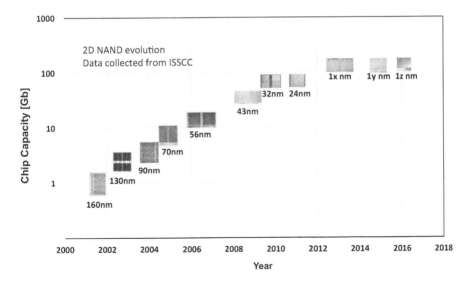

FIGURE 4.38 Historical trend of 2D NAND Flash chip capacity and their technology feature size.

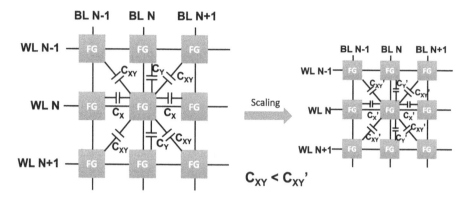

FIGURE 4.39 Schematic of cell-to-cell interference in a NAND Flash array where the floating gate is capacitively coupled to its neighboring cells. The coupling capacitances increase with scaling.

cells are being programmed. Cell-to-cell interference becomes one of the most important scaling challenges.

In the NAND Flash operation, there is a sequence of the programming order page-by-page, as shown in Figure 4.40(a). For example, WL1 has been programmed first, and then WL2 is being programmed. Due to the capacitive coupling, WL1's cell's V_T may further be increased due to the adjacent cell's V_{PGM} partially adding on top of WL1's V_{passW}. Along WL2, even some of the cells are being inhibited, the capacitive coupling may also induce larger voltage than those cells, increasing their V_T. Figure 4.40(b) shows the V_T distribution of an intermediate programmed state in NAND Flash after cell-to-cell interference. First, after the adjacent BL programming, the V_T shifts slightly toward the right; second, after the adjacent WL programming, the V_T shifts further toward the right. This cell-to-cell interference may cause a severe problem in MLC/TLC/QLC operations, as the intermediate programmed states may overlap with each other due to the reduced margin in between.

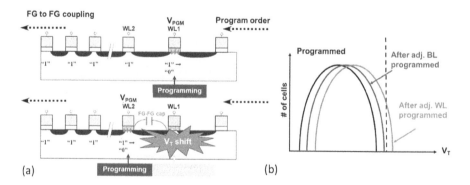

FIGURE 4.40 (a) The page-by-page programming order and the WL1's V_T increases when WL2 is being programmed due to the capacitive coupling. (b) The V_T distribution of an intermediate programmed state in NAND Flash after cell-to-cell interference.

Flash Memory

To mitigate the cell-to-cell interference, the floating-gate device structure engineering has been introduced. First, the height of the floating gate is reduced as shown in Figure 4.41. As a result, the overlapping area (thus the coupling capacitance) between BLs and WLs is reduced. The ultimate floating-gate structure is a planar control gate configuration. A possible trade-off here is the reduced coupling ratio from the control gate to the floating gate due to the smaller inter-poly dielectric area. Hence, it is less effective of the control gate voltage to tune the floating gate voltage, leading to higher V_{PGM}. Second, the air gap is used between the floating gates to reduce the coupling capacitance as the air is approaching the fundamental lower limit of permittivity ~1, as shown in Figure 4.42. The air gap technology has been used in sub-20 nm 2D NAND Flash generations.

4.6.2 Few Electrons Problem

With the aggressive scaling, the number of electrons that are stored in the floating gate becomes less and less simply because of the reduced area of the floating gate. As

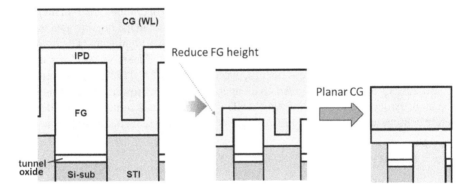

FIGURE 4.41 Floating-gate device structure engineering strategies to reduce capacitive coupling between cells: reduce floating gate height and use planar control gate.

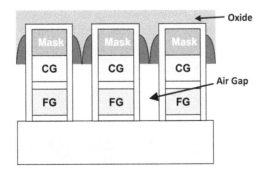

FIGURE 4.42 The air gap introduced between WLs to reduce the capacitive coupling between cells.

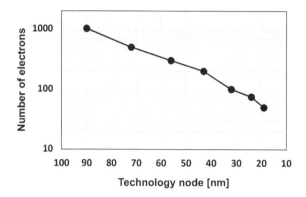

FIGURE 4.43 Scaling trend of the number of electrons of a floating-gate transistor that stores in its fully programmed state to show the few electrons problem.

a result, the data retention becomes vulnerable to any loss of electrons. This is thus referred to as the "few electrons" problem. Figure 4.43 shows the number of electrons of a floating-gate transistor that stores in its fully programmed state. In the sub-20 nm 2D NAND Flash era, the number of electrons stored in the floating gate is less than 50. Therefore, the variation that any electron leaks is significant. Therefore, it is becoming difficult to maintain distinguishable states and memory windows under the data retention, program/read disturb, especially for MLC/TLC/QLC operations.

Because of the aforementioned challenges in 2D scaling toward the 10 nm node, the industry decided to switch to 3D NAND Flash to mitigate some of the challenges while continuously improving the integration density and reducing the bit cost.

4.7 3D NAND FLASH

When the 2D NAND hit the scaling limit around 14 nm feature size in the mid-2010s, the industry adopted the 3D NAND, as shown in Figure 4.44. Though the lateral feature size is relaxed to >100 nm in the first generation of 3D NAND, the scaling trend of the equivalent bit area continues scaling in the 3D NAND era, thanks to the increase of the 3D NAND layers from 24 to more than 128. It is necessary to introduce the charge-trap transistor before discussing the 3D NAND array architecture because it is a commonly used device structure to realize the 3D NAND Flash.

4.7.1 Principle of Charge-Trap Transistor

The operational principle of a charge-trap transistor is similar to that of a floating-gate transistor. The key difference is that the charge-trap transistor uses the nitride-based charge-trap layer to replace the poly-silicon-based floating-gate layer. The nitride layer is able to capture the electrons that are injected from the substrate. Figure 4.45 shows the program/erase mechanism for the charge-trap transistor. The program is similar to the floating-gate transistor with electron injection by the channel F-N tunneling. A notable difference here is the erase mechanism is assisted by

Flash Memory

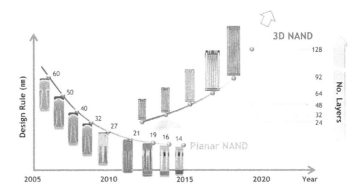

FIGURE 4.44 The transition from the 2D planar NAND to the 3D vertical NAND.

the hole injection from the substrate in addition to the electron ejection. Figure 4.46 shows the evolution of the nitride-based charge-trap transistor. The charge-trap transistor actually has been developed almost in parallel to the floating-gate transistor since the 1970s. A representative structure of the charge-trap transistor that is used in 3D NAND Flash is called MONOS, which stands for the metal/SiO_2/Si_3N_4/SiO_2/(poly-)silicon stack.

The initial motivation for the charge-trap transistor is lower programming voltage (typically sub-10 V) compared to that of the floating-gate transistor (~20 V). Another advantage of the charge-trap transistor is potentially better program/erase cycling endurance and retention. In particular, the charge-trap transistor is immune to the single defect loss of charges. This is because the trapped charges in the nitride layer are not movable. If there is a leakage path to a certain location of the nitride layer, the local charges in that location may leak out while the charges in other locations will be retained. This is unlike the floating-gate transistor where all the charges

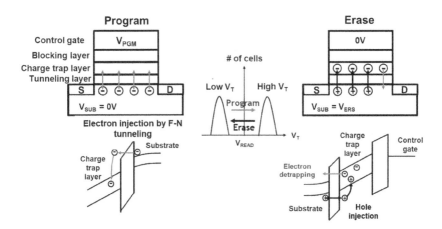

FIGURE 4.45 The program/erase mechanism for the charge-trap transistor.

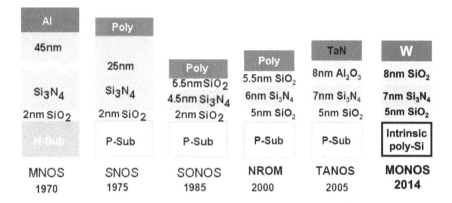

FIGURE 4.46 The evolution of the nitride-based charge-trap transistor. MONOS is used in 3D NAND Flash.

may leak out through a single leakage path as the charges are essentially movable in the poly-silicon layer.

4.7.2 COST-EFFECTIVE 3D INTEGRATION APPROACHES

The aforementioned challenges in the 2D scaling of NAND Flash motivate the transition toward the 3D integration for higher bit density in terms of Gb/mm^2. Nevertheless, improving bit density is not the final goal for the industry as the fabrication cost (in terms of $/mm^2$) is also an important factor to consider. The memory manufacturers will not adopt a 3D integration approach that is costly. What matters ultimately is the bit cost (in terms of $/Gb, defined as fabrication cost divided by bit density). Figure 4.47 shows possible 3D integration approaches for NVMs in general. The first approach is based on the stacked cells. Essentially, it repeats the same fabrication processes layer-by-layer, for instance, creating the simply stacked NAND. This approach is also used for 3D X-point memory.[8] However, this approach does not save the bit cost as the number of lithography steps increases linearly with the number of 3D layers. Generally, lithography patterning (including the subsequent etching) is expensive, and the fabrication cost is proportional to the mask count. Hence, the stacking approach is not attractive for lowering the bit cost. This cost-sensitive nature of NAND Flash also does not prefer the heterogeneous integration in 3D DRAM that uses TSV and hybrid bonding with multiple dies.

Alternatively, the NAND Flash adopts a monolithic 3D integration approach based on vertical cells, which share some common lithography steps either in the channel formation (i.e., vertical channel) or the gate patterning (i.e., vertical gate). The vertical gate 3D NAND is similar to today's gate-all-around stacked nanosheet transistor for the logic process, where multiple planar channels share a common vertical gate. Though demonstrated in a research paper [3], this vertical gate idea did not work out in manufacturing. The industry all embraced the vertical channel 3D NAND, and there are two fabrication processes: the gate-first and the gate-last (similar to the high-k/metal gate process flow options in the logic process). Intel uses the

Flash Memory

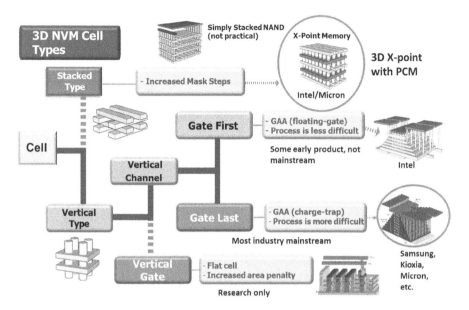

FIGURE 4.47 The general 3D integration approaches for NVMs.

gate-first process that is compatible with the floating-gate transistor, while other vendors including Samsung, SK Hynix, and Kioxia all use the gate-last process that is compatible with the charge-trap transistor. Micron initially followed the same approach as Intel, but later switched to the mainstream by other major vendors. The vertical cell approach could reduce the bit cost as multiple layers share one critical lithography step, for instance, to pattern the vertical string for the channel.

Next, the bit-cost-scalable (BiCS) array architecture is presented to illustrate the essence of vertical channel 3D NAND design. The BiCS concept was first proposed by Toshiba (now Kioxia) at the Symposium on VLSI Technology in 2007 [4], and has become the pioneering and representative design for 3D NAND, inspiring the later designs such as Samsung's Terabit Cell Array Transistor (TCAT) array architecture presented at Symposium on VLSI Technology in 2009 [5]. Figure 4.48 shows the schematic of the BiCS array architecture. The vertical pillar in the memory array core serves as the channel and multiple WL layers surround the pillars to form the gate-all-around structure. One layer shares a common WL terminal at the edge though a WL-cut splits the pillars away. One pillar naturally forms one NAND string as multiple layers are connected in series by the vertical channel. Besides the charge-trap gate stack, another notable feature here is employing the poly-silicon vertical channel (as opposed to the single-crystalline silicon in 2D NAND substrate). The program of the 3D vertical NAND is through the normal F-N tunneling, while the erase is through the GIDL effect near the source followed by hole tunneling into the charge-trap layer. This unique erase mechanism is attributed to the lack of access to the substrate as the vertical pillar is fully depleted. Near the top/bottom end of the string, there is an upper/lower string select layer with a regular dielectric as the gate stack. The upper SSLs are individually addressable, while the lower GSLs are common.

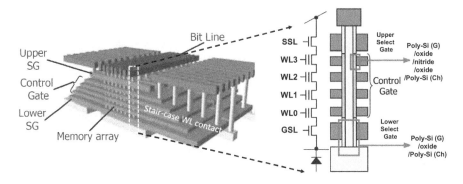

FIGURE 4.48 Schematic of the BiCS array architecture as one of the representative 3D vertical channel NAND Flash designs.

The top end of the string is connected to individual BL, and the bottom end of the string is connected to a common SL. In order to access different layers of WL, a staircase WL contact via region is needed at the edge of the memory array core area. Due to different contact via heights, WL in different 3D layers could be reached.

Figure 4.49 shows the equivalent circuit diagram for the 3D NAND array architecture. An x-y-z decoding scheme is needed for addressing a 3D NAND block. First, the x-direction is decoded by a BL decoder that could select BLs. Second, the y-direction is decoded by the upper SSL, which is able to address individual strings along the same BL. Third, the z-direction is decoded by a WL decoder that could select different layers. Essentially, the x-z plane of a 3D NAND is exactly the same as a 2D NAND. In other words, 3D NAND is an accumulation of x-z slices along the

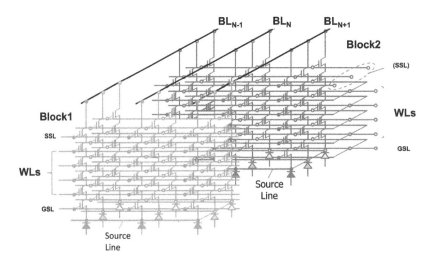

FIGURE 4.49 The equivalent circuit diagram for the 3D NAND array architecture showing two blocks.

y-direction. Typically, a 3D NAND block has a size of 8 kB–16 kB BLs, and only 4–8 SSL, and the 3D vertical layers could range from 24 WLs to 176 WLs (as of 2020).

4.7.3 3D NAND Fabrication Issues

Since most of the vendors adopt the gate-last process for the vertical channel 3D NAND, its generic process flow is discussed here. Figure 4.50 shows the memory array core's representative fabrication flow: (a) multiple interleaved oxide/nitride layers are deposited; (b) the vertical channel region is defined by one lithography step and holes are etched through multiple layers; (c) the oxide/nitride/oxide charge-trap layer and the poly-silicon channel are sequentially deposited covering the sidewall of the hole; (d) the middle trench of the holes are filled with oxide; (e) the WL-cut is done to etch slits; (f) the nitride layer is removed by chemical isotropic etching; (g) the metal is filled into the slits; (h) excessive metal is etched away for the WL-cut.

At the edge of the array, there is a staircase WL contact region that needs a special fabrication treatment. Since different WL layers will be exposed to contact vias separately, the etching needs to be stopped at different stairs. A naïve way is to utilize different lithography masks to pattern different stairs, but this will increase the fabrication cost thus it is not bit-cost-scalable. A creative fabrication method called the photoresist trimming allows only one critical lithography step in the entire staircase region, as shown in Figure 4.51. The lithography defines the shape of the photoresist after the regular exposure and developing process. The topmost layer's stair is etched using the photoresist as the mask. The etching condition is well controlled to selectively stop at the surface of the second layer. Then, the photoresist is slimed isotropically by a trimming process. The selective etching is conducted again to reach the surface of the third layer. Such a process repeats itself until all the stairs are fabricated. Therefore, the WL contact vias could be fabricated by simply etching holes down to

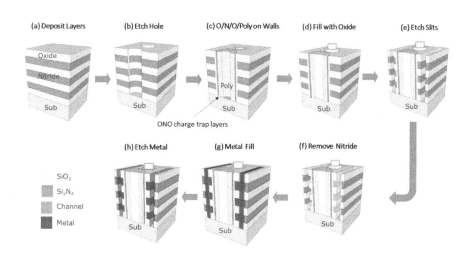

FIGURE 4.50 The representative gate-last fabrication flow for 3D NAND Flash.

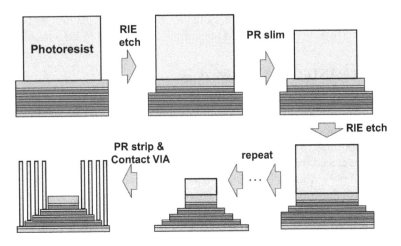

FIGURE 4.51 Photoresist trimming process to fabricate the staircase WL contact region using one critical lithography step.

each step of the staircase region. One obvious drawback of such staircase WL access is a substantial area penalty that increases with the number of WL layers.

Apparently, the limiting factor of 3D NAND fabrication is no longer the advanced lithography (as the hole diameter is relaxed with 60–120 nm feature size). Instead, the high aspect ratio (up to 60) etching through dozens of layers is now the new challenge since the etching profile needs to approach a nearly perfect 90° vertical sidewall. In addition, the atomic layer deposition (ALD) technique is also critical to the conformal growth of the oxide/nitride/metal that uniformly covers the sidewall including the under-etched region in the slits.

4.7.4 Analysis of the First-generation 3D NAND Chip

After a few years of industrial research and development since 2007, Samsung was the first company to commercialize the 3D NAND, where it presented its first generation of 3D NAND chip at the IEEE International Solid-State Circuits Conference (ISSCC) in 2014 [6]. This is a 128-Gb chip with 24 WL layers. There are two planes, and each 64-Gb plane has 2732 blocks, then each block has a page size = 8 kB BLs and 8 SSLs. Figure 4.52 shows the die photo of the first-generation 3D NAND by Samsung and its cross-sectional transmission electron microscopy (TEM) image. From the reverse engineering report [7], the top-view layout shows that a hexagonal vertical pillar is distributed in the memory array core region. The staircase WL access region is placed at the edge of the array. A closer look reveals a complicated gate stack consisting of a multi-tier oxide/nitride charge-trapping layer, TiN, and W gate together with the poly-silicon channel.

Comparing the first-generation 3D NAND is with its predecessor, the 1x nm planar 2D NAND, the bit density increases by a factor of 1.64× and the write bandwidth increases by a factor of 1.5, showing a significant boost. In addition, the memory cell characteristics of this first-generation of 3D NAND demonstrate superior reliability

Flash Memory

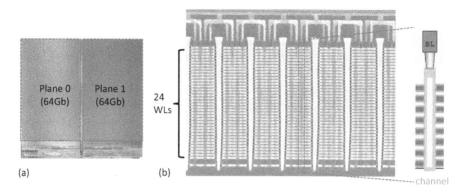

FIGURE 4.52 The die photo of the first-generation 24-layer 3D NAND by Samsung and its cross-sectional TEM image.

than the 2D NAND counterpart at 1x nm node. Figure 4.53(a) shows the comparison between the V_T distribution of the 3D NAND and that of the 2D NAND. V_T of a certain state has 33% less spread-out of its distribution in 3D NAND than in 2D NAND. Cell-to-cell interference is much suppressed in 3D NAND with 84% less V_T shift after adjacent cells' programming compared to 2D NAND. These improvements are partly attributed to the transition from the floating-gate transistor to the charge-trap transistor, and largely attributed to the increased gate area. A simple calculation could be analyzed here, as shown in Figure 4.53(b). For 2D NAND at 1x nm node, i.e., $F = 19$ nm, the gate area is $F^2 = 361$ nm^2, while for 3D NAND, assuming $F = 120$ nm, the gate area is $\pi \times F \times H = 3.14 \times 120$ nm $\times 30$ nm $= 11{,}304$ nm^2, where H is the height of one WL layer ~ 30 nm. Therefore, the 3D NAND has a >31× increase of the gate area, and it is equivalent to the 100 nm node 2D NAND's gate area. The few electrons problem is much mitigated and the cell-to-cell coupling is also reduced. In sum, 3D NAND has much relaxed dimensions; therefore, the reliability is much improved.

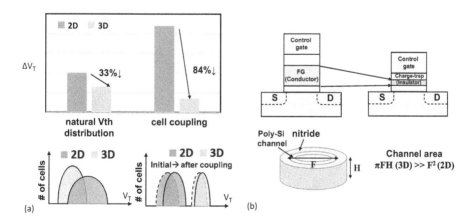

FIGURE 4.53 (a) The comparison between the V_T distribution of the 3D NAND and that of the 2D NAND. (b) The effective channel area in the 3D NAND and in the 2D NAND.

4.7.5 NEW TRENDS IN 3D NAND

Since the first-generation 3D NAND introduced by Samsung in 2014, other vendors started offering 3D NAND in 2016. Then, the number of WL layers increases rapidly each year, from 24 to 48, 72, 96, and 144 to 176 (as of 2020). Figure 4.54 shows the trend of 3D NAND WL layers by different manufacturers. Micron, SK Hynix, and Kioxia introduced the 2-deck string stacking in the vertical pillar around 2018. The 2-deck design is to split the vertical pillar patterning into two critical lithography steps (instead of one) as the number of WLs increases to around 72–96. For example, Micron's 176-layer 3D NAND is built as 88 layers of 3D NAND strings on a silicon substrate, and then another 88 layers aligned on top of the underlying string. This is mainly due to the difficulties in the high aspect ratio etching in one shot for a larger number of layers. Inevitably, the fabrication cost of the 2-deck design is increased. Nevertheless, Samsung has been able to avoid using string stacking and can manufacture 128 layers as a single deck while the other vendors have to split their stack into two decks. Samsung is expected to use the 2-deck design in the 176-layer generation.

The other way to improve the integration density is to move beyond MLC to TLC/QLC. Tables 4.2 and 4.3 summarize the recent TLC and QLC 3D NAND, respectively, that are offered by different vendors as reported in ISSCC 2019 to 2021 [8–14]. TLC 3D NAND chips introduced by SK Hynix and Kioxia look comparable in specifications, except that SK Hynix is offering a 512-Gb die while Kioxia is offering a 1 Tb die.

Intel has been more focused on QLC NAND than other vendors. Intel's 144-layer QLC is the first generation of 3D NAND that Intel did not co-developed with Micron, and it is unique in several aspects. Intel's 144-layer QLC design also introduced a 3-deck stack design: 48 + 48 + 48 layers rather than the 72 + 72 layers. Since Intel's previous generation 96-layer QLC [15] is a 48 + 48 layers design; thus, it is possible that the fabrication steps simply repeated the same sequence of deposition, etch, and fill steps a third time. Intel is taking some overhead on the fabrication cost with this approach, but it probably helps with better control the variation in channel and cell

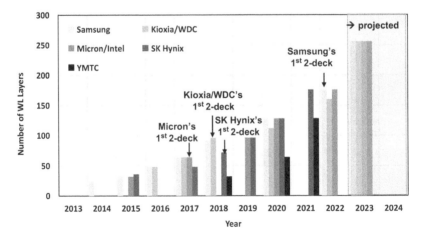

FIGURE 4.54 The trend of 3D NAND WL layers by different manufacturers.

TABLE 4.2
Summary of the Recent TLC 3D NAND as Reported in ISSCC

3D TLC NAND Flash Memory			
	Samsung	SK Hynix	Kioxia/WDC
Year at ISSCC	2021	2021	2021
Layers	>170	176	>170
Die Capacity	512 Gb	512 Gb	1 Tb
Die Size (mm²)	60.2	47.4	98
Density (Gbit/mm²)	8.5	10.8	10.4
I/O Bandwidth	2.0 Gb/s	1.6 Gb/s	2.0 Gb/s
Program Throughput	184 MB/s	168 MB/s	160 MB/s
Program Latency (tPROG)	350 µs	380 µs	400 µs
Read Latency (tR)	40 µs	50 µs	50 µs
Planes	4	4	4
CuA/PuC	Yes	Yes	Yes

TABLE 4.3
Summary of Recent QLC 3D NAND as Reported in ISSCC

3D QLC NAND Flash Memory				
	Intel	Samsung	SK Hynix	Kioxia/WDC
Year at ISSCC	2021	2020	2020	2019
Layers	144	92	96	96
Die Capacity	1 Tb	1 Tb	1 Tb	1.33 Tb
Die Size (mm²)	74.0	136	122	158.4
Density (Gbit/mm²)	13.8	7.53	8.4	8.5
I/O Bandwidth	1.2 Gb/s	1.2 Gb/s	800 Mb/s	800 Mb/s
Program Throughput	40 MB/s	18 MB/s	30 MB/s	9.3 MB/s
Program Latency (tPROG)	1630 µs	2 ms	2.15 ms	3380 µs
Read Latency Avg (tR)	85 µs	110 µs	170 µs	160 µs
Planes	4	2	4	2
CuA/PuC	Yes	No	Yes	No

dimensions from the top to the bottom of the stack, which may be more of a concern given their focus on QLC and their unique decision to use a floating-gate transistor rather than a charge-trap transistor.

The most significant innovation in the Intel/Micron 3D NAND joint development era (2015–2019) was the CMOS under the Array (CuA) concept.[9] As shown in Figure 4.55(a), CuA places most of the NAND die's peripheral circuitry, including page buffers, sense amplifiers, charge pumps, etc., under the vertical stack of memory cells instead of alongside. It should be noted that CuA still employs a monolithic 3D fabrication process that sequentially fabricates CMOS and NAND on the same

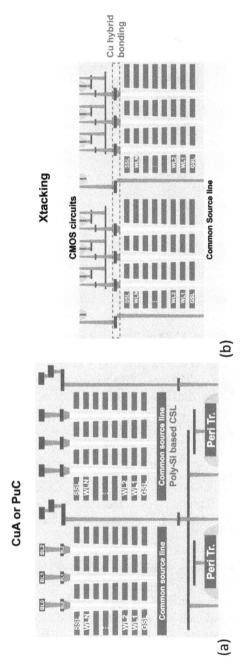

FIGURE 4.55 (a) Schematic of CMOS Under the Array (CuA) of 3D NAND. (b) Schematic of X-stacking that uses hybrid bonding of a NAND die and a CMOS die.

substrate. The newcomer to 3D NAND is YMTC, a vendor that advocates the X-stacking that uses hybrid bonding of a NAND die and a CMOS die, as shown in Figure 4.55(b).

With regard to future 3D NAND development and integration density improvement, it is expecting a combination of approaches such as more string stacking (e.g., 3-deck or 4-deck), higher memory cell precision (moving from TLC to PLC, Penta-Level Cell), CuA, and/or hybrid bonding.

NOTES

1. V_{T_L} could be a negative value for NAND Flash.
2. Technically, it should be drain F-N tunneling as a high voltage is applied to NMOS transistor's drain; however, by convention, it is referred to as source F-N tunneling.
3. The injection occurs at the lower potential side of the floating-gate transistor, and thus it is referred to as the source-side injection.
4. 2D NAND is discussed in this section, while 3D NAND is discussed in Section 4.7. Many operation principles of 2D NAND could be applied to 3D NAND.
5. Strictly speaking, MLC is referred to as 2 bits per cell. In a more general definition, MLC is referred to all the multilevel cells, including 2 bits per cell, 3 bits per cell, and 4 bits per cell.
6. k is the Boltzmann constant. kT is the thermal energy.
7. Vendors generally do not disclose the exact feature size for the sub-20 nm generations, 1x is roughly 19 nm, 1y is roughly 17 nm, and 1z is roughly 14 nm.
8. 3D X-point memory utilizes the phase change memory to be discussed in Section 5.2.
9. CuA is also called Periphery under Cell (PuC) by other vendors.

REFERENCES

[1] F. Masuoka, M. Asano, H. Iwahashi, T. Komuro, S. Tanaka, "A new flash E²PROM cell using triple polysilicon technology," *IEEE International Electron Devices Meeting (IEDM)*, 1984, pp. 464–467. doi: 10.1109/IEDM.1984.190752

[2] F. Masuoka, M. Momodomi, Y. Iwata, R. Shirota, "New ultra high density EPROM and flash EEPROM with NAND structure cell," *IEEE International Electron Devices Meeting (IEDM)*, 1987, pp. 552–555. doi: 10.1109/IEDM.1987.191485

[3] H.-T. Lue, T.-H. Hsu, Y.-H. Hsiao, S.P. Hong, M.T. Wu, F.H. Hsu, N.Z. Lien, et al., "A highly scalable 8-layer 3D vertical-gate (VG) TFT NAND flash using junction-free buried channel BE-SONOS device," *IEEE Symposium on VLSI Technology*, 2010, pp. 131–132. doi: 10.1109/VLSIT.2010.5556199

[4] H. Tanaka, M. Kido, K. Yahashi, M. Oomura, R. Katsumata, M. Kito, Y. Fukuzumi, et al., "Bit cost scalable technology with punch and plug process for ultra high density flash memory," *IEEE Symposium on VLSI Technology*, 2007, pp. 14–15. doi: 10.1109/VLSIT.2007.4339708

[5] J. Jang, H.-S. Kim, W. Cho, H. Cho, J. Kim, S.I. Shim, J.-H. Jeong, et al., "Vertical cell array using TCAT (Terabit Cell Array Transistor) technology for ultra high density NAND flash memory," *IEEE Symposium on VLSI Technology*, 2009, pp. 192–193.

[6] K.-T. Park, J.-M. Han, D. Kim, S. Nam, K. Choi, M.-S. Kim, P. Kwak, et al., "Three-dimensional 128Gb MLC vertical NAND Flash-memory with 24-WL stacked layers and 50MB/s high-speed programming," *IEEE International Solid-State Circuits Conference (ISSCC)*, 2014, pp. 334–335. doi: 10.1109/ISSCC.2014.6757458

[7] Reverse engineering report on Samsung's 1st generation 3D NAND chip (by ChipWorks), http://chipworksrealchips.blogspot.com/2014/08/the-second-shoe-drops-now-we-have.html
[8] J. Cho, D.C. Kang, J. Park, S.-W. Nam, J.-H. Song, B.-K. Jung, et al., "A 512Gb 3b/cell 7th-generation 3D-NAND Flash memory with 184MB/s write throughput and 2.0Gb/s interface," *IEEE International Solid-State Circuits Conference (ISSCC)*, 2021, pp. 426–428. doi: 10.1109/ISSCC42613.2021.9366054
[9] J.-W. Park, D. Kim, S. Ok, J. Park, T. Kwon, H. Lee, S. Lim, et al., "A 176-stacked 512Gb 3b/cell 3D-NAND Flash with 10.8Gb/mm^2 density with a peripheral circuit under cell array architecture," *IEEE International Solid-State Circuits Conference (ISSCC)*, 2021, pp. 422–423. doi: 10.1109/ISSCC42613.2021.9365809
[10] T. Higuchi, T. Kodama, K. Kato, R. Fukuda, N. Tokiwa, M. Abe, T. Takagiwa, et al., "A 1Tb 3b/cell 3D-Flash memory in a 170+ word-line-layer technology," *IEEE International Solid-State Circuits Conference (ISSCC)*, 2021, pp. 428–430. doi: 10.1109/ISSCC42613.2021.9366003
[11] A. Khakifirooz, S. Balasubrahmanyam, R. Fastow, K.H. Gaewsky, C.W. Ha, R. Haque, et al., "A 1Tb 4b/cell 144-tier floating-gate 3D-NAND Flash memory with 40MB/s program throughput and 13.8Gb/mm^2 bit density," *IEEE International Solid-State Circuits Conference (ISSCC)*, 2021, pp. 424–426. doi: 10.1109/ISSCC42613.2021.9365777
[12] D. Kim, H. Kim, S. Yun, Y. Song, J. Kim, S.-M. Joe, K.-H. Kang, J. Jang, et al., "A 1Tb 4b/cell NAND Flash memory with t_{PROG}=2ms, t_R=110μs and 1.2Gb/s high-speed IO rate," *IEEE International Solid-State Circuits Conference (ISSCC)*, 2020, pp. 218–220. doi: 10.1109/ISSCC19947.2020.9063053
[13] H. Huh, W. Cho, J. Lee, Y. Noh, Y. Park, S. Ok, J. Kim, K. Cho, et al., "A 1Tb 4b/cell 96-stacked-WL 3D NAND Flash memory with 30MB/s program throughput using peripheral circuit under memory cell array technique," *IEEE International Solid-State Circuits Conference (ISSCC)*, 2020, pp. 220–221. doi: 10.1109/ISSCC19947.2020.9063117
[14] N. Shibata, K. Kanda, T. Shimizu, J. Nakai, O. Nagao, N. Kobayashi, M. Miakashi, Y. Nagadomi, T. Nakano, et al., "A 1.33Tb 4-bit/cell 3D-Flash memory on a 96-word-line-layer technology," *IEEE International Solid-State Circuits Conference (ISSCC)*, 2019, pp. 210–212. doi: 10.1109/ISSCC.2019.8662443
[15] P. Kalavade, "4 bits/cell 96 layer floating gate 3D NAND with CMOS under array technology and SSDs," *IEEE International Memory Workshop (IMW)*, 2020, pp. 1–4. doi: 10.1109/IMW48823.2020.9108135

5 Emerging Non-volatile Memories

5.1 ENVM OVERVIEW

5.1.1 Landscape of eNVMs

Though mainstream memory technologies (SRAM, DRAM, and Flash) dominate the market today, emerging non-volatile memory (eNVM) technologies have made significant progress in research and development since 2010. Figure 5.1 shows a simplified taxonomy of memory technologies. First, mainstream technologies are all charge-based memories. For example, SRAM stores the charges by parasitic capacitance in the storage nodes of the cross-coupled inverters. DRAM stores the charges by an explicit 3D cylinder-type capacitor. Flash stores the charges in the floating-gate or the charge-trapping layer. Second, the emerging technologies could be classified into two types: one is the resistive memories,[1] the other is the capacitive memories.

The resistive memories include phase change memory (PCM), resistive random access memory (RRAM), spin-transfer-torque magnetic random access memory (STT-MRAM), and spin-orbit-toque magnetic random access memory (SOT-MRAM). These resistive memories share some common features: they are non-volatile two-terminal devices,[2] and they differentiate their memory states by switching between a high resistance state (HRS, or off-state) and a low resistance state (LRS, or on-state). The switching from off-state to on-state is called "set", and the switching from on-state to off-state is called "reset". The transition between the two states can be triggered by electrical stimulus (i.e., voltage or current pulse). However, the detailed switching physics is quite different: PCM relies on chalcogenide materials to switch between the crystalline phase (corresponding to LRS) and the amorphous phase (corresponding to HRS); RRAM relies on the formation (corresponding to LRS) and the rupture (corresponding to HRS) of conductive filaments in the insulator between two electrodes, and MRAM relies on the parallel magnetization orientation (corresponding to LRS) and anti-parallel magnetization orientation (corresponding to HRS) of two ferromagnetic layers separated by a thin tunneling insulator layer.

The capacitive memories are mostly ferroelectric memories, where the different polarities of the spontaneous polarization in ferroelectric materials are used to represent the memory states, which generally reflect a change in the capacitance value in the device.[3] Ferroelectric random access memory (FeRAM) employs a similar 1T1C bit cell as DRAM, where the ferroelectric capacitor with different polarization states

DOI: 10.1201/9781003138747-5

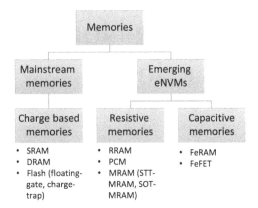

FIGURE 5.1 Simplified taxonomy of the memory technologies.

will induce different amounts of charges when being read out. The ferroelectric field-effect transistor (FeFET) integrates the ferroelectric layer into the gate stack, where the ferroelectric capacitor with different polarization states will facilitate or inhibit the channel formation underneath, thereby modulating the threshold voltage (similar to the Flash transistor).

Table 5.1 shows the comparison of the key attributes between mainstream memories and eNVMs. Most of eNVMs have a cell area ranging from 4 F^2 to 100 F^2. Some of the eNVMs are capable of multilevel storage, for example, 2–3 bits per cell. SRAM and DRAM have low operation voltage (<1 V), while Flash has high write voltage (>10 V). Generally, eNVMs have moderate write voltage (2–3 V) except for MRAM (with <1 V). Moderate write voltage eases the overhead of large charge pump circuitry but still some sort of level shifter is necessary for eNVMs, because 2–3 V is still higher than the logic power supply voltage. In terms of access speed, SRAM is the fastest and is always difficult to compete with. eNVMs generally achieve read/write time (~10 ns), which is comparable to NOR Flash for the read but much faster for the write. eNVMs are difficult to realize sub-ns write operation but may approach DRAM's range in a few nanoseconds if optimized. Retention and endurance may have trade-offs. Typical year-long data retention is possible, but the endurance is limited to 10^6–10^9 cycles for most eNVMs. MRAM, if relaxing the data retention requirement, may reach 10^{12}–10^{14} cycles. In terms of the write energy at the bit cell level, SRAM charging/discharging energy is super low (~fJ) because of the sub-fF parasitic capacitance of the storage node. DRAM charging/discharging energy is higher as it requires an explicit ~10 fJ capacitance for maintaining sufficient sense margin. NOR Flash has a very high write energy (>100 pJ) due to the channel hot electron programming's low efficiency, while NAND Flash has a relatively low write energy due to the electric field-driven FN tunneling mechanism (but it still needs a long programming/erase time). PCM has relatively high write energy (>10 pJ) as it needs to melt the material by raising the temperature. RRAM and MRAM have moderate write energy (~pJ) due to the current-driven switching mechanism. FeRAM has a slightly larger energy consumption than DRAM because it needs to flip the polarization in addition to the

Emerging Non-volatile Memories

TABLE 5.1
The Comparison of the Key Attributes between Mainstream Memories and eNVMs

	Mainstream Charged based Memories				Emerging Non-volatile Memories				
			FLASH						
	SRAM	DRAM	NOR	NAND	PCM	RRAM	STT-MRAM	SOT-MRAM	FeRAM	FeFET
Cell area	>150 F^2	6 F^2	10 F^2	<4 F^2 (3D)	4–50 F^2	4–50 F^2	6–50 F^2	12–100 F^2	6–50 F^2	6–50 F^2
Multi-bit	1	1	2	3–4	2–3	2–3	1	1	1	2–3
Voltage	<1 V	<1 V	>10 V	>10 V	<3 V	<3 V	<1 V	<1 V	<2 V	<3 V
Read time	~1 ns	~10 ns	~50 ns	~10 μs	<10 ns	<10 ns	<10 ns	~1 ns	<100 ns	<50 ns
Write time	~1 ns	~10 ns	10 μs–1 ms	100 μs–1 ms	~50 ns	<100 ns	<20 ns	<3 ns	<100 ns	<100 ns
Retention	N/A	~64 ms	>10 y	>10 y	>10 y	>10 y	>1 y	>1 y	>10 y	>1 y
Endurance	>1E16	>1E16	~1E5	1E3~1E4	1E6~1E9	1E3~1E9	1E6~1E14	~1E12	1E9~1E12	1E6~1E9
Write Energy (J/bit)	~fJ	~10 fJ	100J	~10 fJ	~10 pJ	~pJ	~pJ	~pJ	~100 fJ	~fJ

F: feature size of the lithography.
The energy estimation is on the cell-level (not on the array-level).
PCM/RRAM/FeFET can potentially achieve less than 4F^2 through 3D integration.
The numbers of this table are representative (not the best or the worst cases).

charging/discharging effect. FeFET has the lowest write energy (~fJ) among all the eNVMs and approaches SRAM's level because it is purely electric-field-driven and the switching is relatively fast (<100 ns).

Figures 5.2 and 5.3 show the recent trends of the eNVM prototype chips (reported in the major conferences) in terms of capacity (in Mb) and density (in MB/mm^2), respectively. State-of-the-art eNVMs' capacity and density are still in between SRAM and DRAM, and 3D NAND is clearly a winner in both capacity and density. It is challenging for eNVMs to compete with 3D NAND in terms of the bit cost. Apparently, eNVMs hold capacity/density advantages over SRAM. Nevertheless, to be more competitive over the DRAM, the 3D integration of eNVMs is a must.

Figure 5.4 shows the recent trend of the eNVM prototype chips (reported in the major conferences) in terms of the read bandwidth and the write bandwidth (in MB/s). eNVMs generally outperform NOR/NAND Flash by a few times to 10 times in terms of the read/write bandwidth, but they are lagging behind substantially than the DRAM families (DDR5, LPDDR5, GDDR6, HBM3, etc.).

Apparently, there is no universal memory that could satisfy all the desired attributes. Considering all the pros and cons, the application spaces of eNVMs are mostly in the embedded memories (instead of the standalone memories): (1) replacing the NOR Flash as embedded NVM for microcontrollers (especially at advanced technology nodes below 28 nm), and all types of eNVMs are competitive candidates for this application; (2) substituting eDRAM for the last level cache, and STT-MRAM (if well optimized) and SOT-MRAM are the leading candidates considering the endurance requirement; (3) enabling new applications beyond data storage for in-memory computing (to be discussed in Section 5.6). To summarize, mainstream technologies such as SRAM, DRAM, and NAND Flash are still not replaceable in the memory hierarchy from either the performance or the bit-cost perspective.

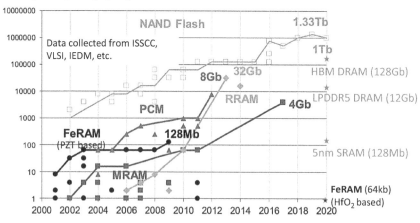

FIGURE 5.2 The recent trend of the eNVM prototype chips in terms of capacity (in Mb). NAND Flash/DRAM/SRAM are also shown as a comparison.

Emerging Non-volatile Memories 137

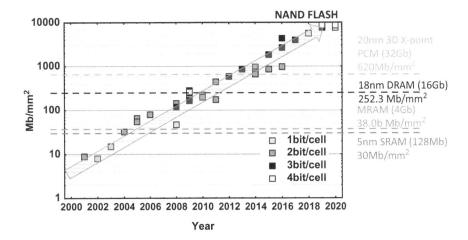

FIGURE 5.3 The recent trend of the NAND Flash density (in MB/mm²). SRAM/DRAM and MRAM/PCM (as of 2020) are also shown as a comparison.

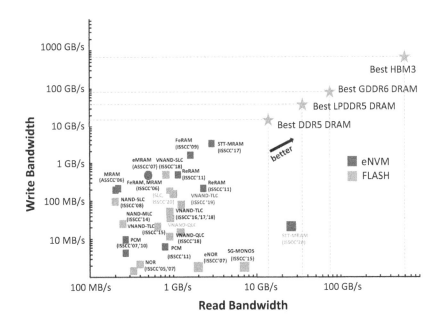

FIGURE 5.4 The recent trend of the eNVM prototype chips in terms of the read bandwidth and write bandwidth (in MB/s). NAND and DRAM with different interface protocols are shown as a comparison.

5.1.2 1T1R Array

While FeRAM and FeFET follow principles similar to those of DRAM and Flash, resistive memories (PCM, RRAM, and STT-MRAM) need new integration strategies, as they are essentially two-terminal variable resistors. To integrate these resistive memories into memory arrays, there are two types of array architectures. The first type is one-transistor and one-resistor (1T1R) array, where each eNVM cell is in series with a cell selection transistor; the second type is the transistor-less crossbar array (to be discussed in Section 5.1.3).

Figure 5.5 shows the 1T1R array architecture and the bit cell structure. Figure 5.5(a)–(c) shows the set/reset/read operation bias, respectively. The selected cell is activated by its WL, and the BL/SL/BL is biased at a set voltage/reset voltage/read voltage, respectively. The BL/SL's role will be exchanged in the bipolar switching devices (i.e., RRAM and STT-MRAM), where the different voltage polarities are required to set or reset. As shown in Figure 5.5(d), the resistive element is integrated at one contact via of the transistor between metal layers. The integration could be done as low as M1/M2 via or as high as M5/M6 via. The integration at lower metal layers could help reduce the cell layout area, and integration at higher metal layers could ease the back-end-of-line (BEOL) fabrication complexity. The addition of a selection transistor could isolate the selected cell from other unselected cells in the array. The WL controls the gate of the transistor; thus, tuning the WL voltage can control the write current that is delivered to the eNVM cell. The eNVM cell's top electrode connects to the BL while its bottom electrode connects to the contact via of the transistor. The source line (SL) connects to the source of the selection transistor. Typically, BL and SL are in parallel, and they are perpendicular to WL. The SL contact could be shared by the two adjacent cells in practice. Essentially, the 1T1R array is similar to the 1T1C array in DRAM if the capacitor is replaced with the resistor.

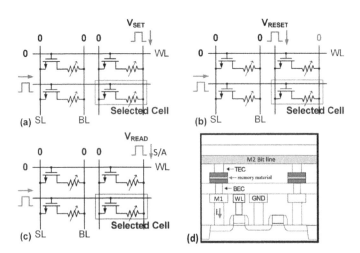

FIGURE 5.5 The 1T1R bit-cell structure and the corresponding array architecture: (a)–(c) the set/reset/read operation bias, respectively; (d) cross-section schematic showing the resistive element is integrated at one contact via of the transistor between metal layers.

Emerging Non-volatile Memories

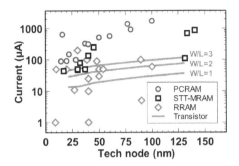

FIGURE 5.6 The survey of PCM/RRAM/STT-MRAM write current and the drivability of the logic transistor with respect to the scaling.

The typical cell area of 1T1R array is ~10 F^2 (F is the lithography feature size) if the gate width/length (W/L) of the transistor is 1. The minimum cell area can be reduced to 6 F^2 if the aggressive borderless DRAM design rule with sharing BL and SL is applied. In practice, the cell area can be 10–50 F^2, because the W/L of the transistor has to be increased in order to provide sufficient write current. Figure 5.6 shows the survey of PCM/RRAM/STT-MRAM write current and the drivability of the logic transistor with respect to the scaling [1]. It shows that PCM and STT-MRAM's write current requirement scales with the technology node; however, their write current is still higher than the contemporary logic transistor with relaxed W/L = 3. Though RRAM exhibits scattered write current due to different stack materials choices, most of the scaled cells need relaxed W/L of the selection transistor.

5.1.3 Cross-Point Array and Selector

The other commonly used array architecture is the cross-point (X-point, or crossbar) array, which consists of rows and columns perpendicular to each other with eNVM cells sandwiched in between without selection transistors. The cross-point array in principle can achieve a 4 F^2 cell area; thus, it can have a higher integration density than the 1T1R array. Typically, the selectors with strong I–V nonlinearity are added in series with the eNVM cells to prevent the cross-talk or interference between cells in the cross-point array, which is referred to as one-selector and one-resistor (1S1R) architecture, as shown in Figure 5.7. The 1S1R array suits PCM and RRAM, but is not well fitted to STT-MRAM due to its small on/off ratio that may be diminished considering the IR drop and sneak path. Figure 5.8 shows the I–V characteristics of the 1S1R cell and the typical definition of the I–V nonlinearity (N), which is the current ratio between the write voltage and the half write voltage for the on-state or LRS.

Two common write schemes (V/2 scheme and V/3 scheme) can be applied to the cross-point array. Figure 5.9(a) shows the voltage bias conditions for the V/2 scheme. In the V/2 scheme, for the set operation, the selected cell's WL and BL are biased at the write voltage V_w and ground, respectively. For the reset operation, the bias conditions on WL and BL are reversed for the bipolar switching. In both set and reset operations, all the unselected WLs and BLs are biased at $V_w/2$. Therefore, only the

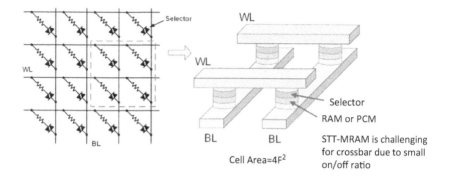

FIGURE 5.7 Cross-point array with one-selector and one-resistor (1S1R) structure.

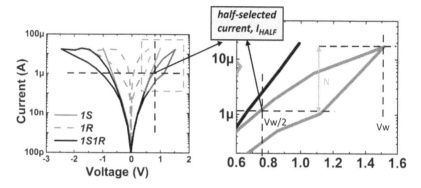

Nonlinearity Ratio (N)=Current(@Vw)/Current(@Vw/2) for V/2 write scheme

FIGURE 5.8 Example of I–V characteristics of the 1S1R cell and the typical definition of the I–V nonlinearity (N) under the V/2 scheme.

selected cell sees a full V_w, while the half-selected cells along the selected WL or BL see a half V_w and all the other unselected cells in the array see zero voltage if wire resistance is neglected. Here, the assumption is that $V_w/2$ should not disturb the half-selected cell's resistance. Figure 5.9(b) shows the voltage bias conditions for the V/3 scheme. In the V/3 scheme, for the set operation, the selected cell's WL and BL are biased at the write voltage V_w and the ground, respectively. For the reset operation, the bias conditions on WL and BL are reversed for the bipolar switching. The unselected WLs and BLs are biased at 1/3 V_w and 2/3 V_w for the set operation, respectively. The unselected WLs and BLs are biased at 2/3 V_w and 1/3 V_w for the reset operation, respectively. In this way, the selected cell sees V_w, while all other unselected cells in the array only see 1/3 V_w. Here, the assumption relaxes to that 1/3 V_w should not disturb the unselected cell's resistance. The pros and cons of these two write schemes can be summarized as follows: the V/2 scheme typically has less power consumption than the V/3 scheme. This is because the unselected cells (not along the selected WL and BL) in the V/2 scheme see zero voltage ideally, while all the unselected cells in the V/3 scheme see 1/3 V_w; thus, consuming static power during

Emerging Non-volatile Memories 141

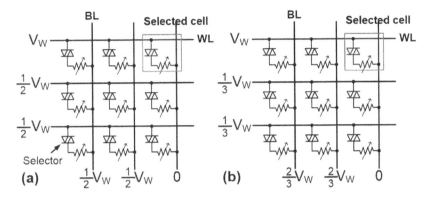

FIGURE 5.9 Write schemes for cross-point array: (a) The voltage bias conditions for the V/2 scheme. (b) The voltage bias conditions for the V/3 scheme.

the write window if the I–V nonlinearity of the selector is insufficient. On the other hand, the V/3 scheme has better immunity to the write disturb than the V/2 scheme, as the maximum voltage that the unselected cells see is 1/3 V_w in the V/3 scheme while is 1/2 V_w in the V/2 scheme. It is possible to have multiple-bit parallel write in the cross-point array with either V/2 or V/3 scheme by biasing multiple BLs (or WLs) to be ground in the set (or reset) operation. The penalty for multiple-bit parallel write is a larger driver size at each row (or column) as it has to deliver the write current to more selected cells in addition to the sneak path via the unselected cells.

The cross-point array suffers from two well-known design challenges: (1) the IR drop problem along the interconnect wire and (2) the sneak path problem through the unselected cells, as shown in Figure 5.10(a). The IR drop problem becomes significant when the WL and BL wire width scales to sub-50 nm regime where the interconnect resistivity drastically increases due to the increased electron surface scattering. For example, at 20 nm node, the copper interconnect resistance between two

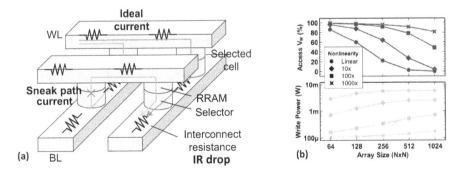

FIGURE 5.10 (a) Design challenges of the cross-point array: (1) the IR drop problem along the interconnect wire and (2) the sneak path problem through the unselected cells. (b) Simulated write voltage access margin (for the farthest cell from the driver) and the write power (of the entire array) as a function of the cross-point array size for different nonlinearity factors (N).

neighboring cells is ~2.93 Ω; thus, the IR drop along the wire for a large array (e.g., 1024 × 1024 array) is no longer negligible. The farthest cell from the driver sees an interconnect resistance of ~3 kΩ. If the eNVM cell's LRS resistance (typically a few kΩ up to tens of kΩ) is comparable to this interconnect resistance, a significant portion of the write voltage will drop on the wire instead of the eNVM cell. To guarantee a successful write operation, the write voltage provided from the driver has to be boosted over the actual switching voltage of the eNVM cell to compensate for the IR drop. However, the write voltage cannot be boosted too much because 1/2 V_w (in the V/2 scheme) should not disturb the eNVM resistance for the cells close to the driver.

The sneak path problem is associated with the IR drop problem. Take the V/2 scheme as an example; the half-selected cells along the selected WL and BL form the sneak paths during the write operation. The sneak paths contribute additional current to the IR drop and further degrade the write margin. Meanwhile, the sneak paths increase the write current (thus the write power) that is provided by the driver transistors at the edge of the cross-point array. It is suggested that increasing the LRS resistance (or equivalently reducing the write current) and increasing the I–V nonlinearity of the eNVM cell (with the help of the selector) are useful to minimize the IR drop and sneak paths.

In order to show the benefits of introducing the selector devices into the cross-point array, an array-level simulation is performed considering the interconnect resistance and sneak paths for the V/2 scheme at 20 nm node [1]. The eNVM cell resistance is assumed to have a low resistance state (LRS) = 40 kΩ, and the worst-case data pattern for the write voltage degradation is that all the cells are at LRS. The nonlinearity (N) of the selector is defined as the current ratio between V_w and $V_w/2$. The write voltage access margin is defined as the voltage dropped on the selected cell over the applied voltage from the peripheral driver. Figure 5.10(b) shows the SPICE simulation results of the write voltage access margin (for the farthest cell from the driver) and the write power (of the entire array) as a function of the cross-point array size for different N values. It is seen that at least N > 1000 is needed for maintaining sufficient write margin and minimizing write power for a large array (i.e., 1024 × 1024 array).

Two common read schemes (current sensing and voltage sensing) can be applied to the cross-point array, as shown in Figure 5.11. For the current sensing, the selected row is biased at the read voltage V_R, while all other unselected rows are biased at zero. All the selected columns are virtually grounded to the input of the sense amplifiers, while the rest unselected columns are floating as a result of the turned-off MUX. For the voltage sensing, the selected row is biased at zero and the unselected rows are biased at V_R. All the selected columns are pre-charged at the read voltage V_R, while the rest unselected columns are floating as a result of the turned-off MUX. Therefore, only the cells of the selected row and the selected columns see a read voltage and all the other unselected cells see zero voltage if ignoring the wire resistance. The selected cells can be read out in parallel by a group of sense amplifiers.

In the following, the voltage sensing scheme is used to illustrate the sneak path problem during the read operation. To properly evaluate the read margin, the worst-case scenario is considered for both cases (reading HRS "0" and reading LRS "1"). As shown in Figure 5.12(a), the worst case for reading of "1" happens when all the cells are in the LRS. In the read operation, as BL voltage decays from pre-charged V_R due

Emerging Non-volatile Memories 143

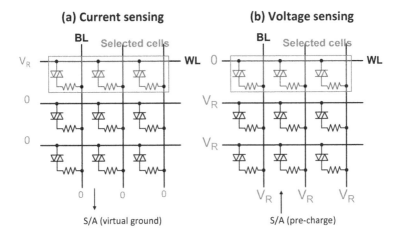

FIGURE 5.11 Read schemes for cross-point array: (a) current sensing and (b) voltage sensing.

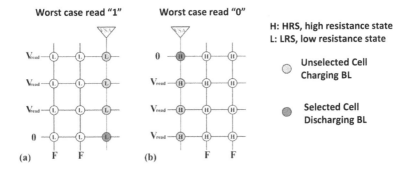

FIGURE 5.12 Schematic of the biasing condition for cross-point array in reading operation with voltage-mode sense amplifier. (a) Worst case for reading LRS "1". (b) Worst case for reading HRS "0".

to the selected cell's current, the leakage current from the unselected rows that are biased at V_R hinders the reading as it charges the BL, preventing the BL fall further. For this reason, the worst case happens when the cell is farthest from the sense amplifier as the voltage on BL will be slightly reduced resulting in less current for the cell being read. As shown in Figure 5.12(b), the worst case for reading of "0" happens when all the cells are in HRS and the cell being read is closest to the sense amplifier. It is the worst case because the BL is receiving the least amount of leakage current to hold the BL voltage high, and there is the least amount of voltage drop due to resistance in the wires when the cell being read is closest to the sense amplifier. This makes it the maximum current, assisting the BL undesired decay.

The floating nature of the unselected columns exacerbates the sneak path problem, as the leakage can flow randomly depending on the exact data pattern. Adding selectors is an effective solution to suppress leakage currents as selectors exhibit nonlinear I–V characteristics that allow them to suppress unwanted currents when low voltages are applied to the unselected cells. Similar to the write access margin analysis above, to determine the relationship between nonlinearity and sensing latency for different array sizes, the cross-point array is simulated using SPICE in 20 nm node considering the interconnect resistance [2]. Assumedly, HRS = 1 MΩ and LRS = 40 kΩ. In the simulation, a 200 mV minimum voltage difference between reading "1" and "0" in the worst cases is set as a sensing success criterion. If this minimum voltage difference is achievable with a certain BL voltage development time, a reference voltage is placed in the middle to split the two states, and the array is considered to be readable. If the minimum voltage difference is not achievable, then the array is considered unreadable. Figure 5.13 shows the BL voltage development time to achieve this minimum voltage difference in different array sizes for different nonlinearity values N from 10 to 1000 (N/A means not achievable). It shows that when nonlinearity is 10, it cannot even create a 200 mV BL voltage difference even for a small 64 × 64 array size. For a fixed array size, a larger nonlinearity will help create the minimum voltage difference faster, thus making read operation faster. For a fixed nonlinearity, when the array size becomes larger, more time is needed to create the minimum voltage difference; in that case, the read operation will become slower.

Figure 5.14 shows a simplified taxonomy of the selector technologies. p/n diode is the unidirectional selector with rectifying I–V, and thus only suitable for PCM. There are two types of bidirectional selectors for both PCM and RRAM: selector with exponential I–V and selector with threshold I–V. Figure 5.15(a)–(b) shows the representative I–V characteristics for a bipolar RRAM integrated with exponential I–V selector, and a bipolar RRAM integrated with threshold I–V selector, respectively. The metrics of selector performance are (1) the nonlinearity (N) defined as the current ratio between V_w and $V_w/2$, which will determine how effective the sneak

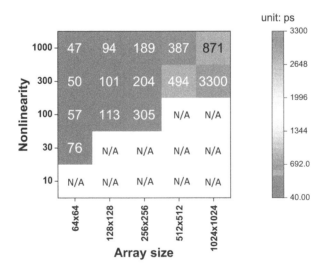

FIGURE 5.13 Simulated BL voltage development time (in unit of ps) for different array sizes and selector nonlinearity values.

Emerging Non-volatile Memories 145

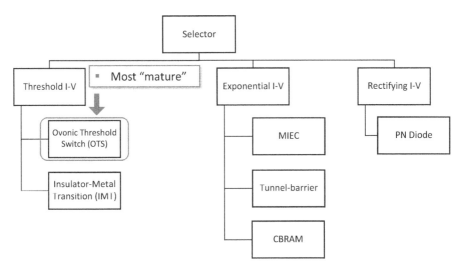

FIGURE 5.14 Simplified taxonomy of the selector technologies.

current suppression is (e.g., >1000 for 1024 × 1024 array); (2) the drive current density (e.g., >12.5 MA/cm^2 required for driving 50 µA write current at the 20 nm node); (3) the cycling endurance (e.g., >10^{12} as the read operation also consumes the cycling endurance for the selector).

Table 5.2 surveys the representative selector devices reported in the literature. Selectors that exhibit an exponential slope in the I–V curve typically rely on the tunneling mechanism. Oxide/electrode interface engineering or oxide/oxide bandgap engineering is thus employed. Examples include Ni/TiO$_2$/Ni selector [3], Pt/TaO$_x$/TiO$_2$/TaO$_x$/Pt selector [4], TiN/a-Si/TiN selector [5], and Ru/TaO$_x$/W bidirectional selector [6]. In addition, Cu ion motion in the Cu-containing Mixed-Ionic-Electronic-Conduction (MIEC) materials [7] also shows a good bidirectional exponential I–V for bipolar switching RRAM.

Selectors that exhibit an abrupt threshold switching are accompanied by the I–V hysteresis, which means the selectors turn on above a threshold voltage and turn off below a hold voltage. Threshold switching can be achieved in insulator-metal-transition (IMT) materials such as NbO$_2$ [8]. The shortcoming of IMT-based threshold selectors is a relatively small nonlinearity (typically N < 100). Besides IMT materials, Ovonic-Threshold-Switch (OTS) [9] based on doped chalcogenide materials have been demonstrated to be a promising threshold selector. Examples of OTS include complex chalcogenides (e.g., TeAsGeSiN [10]) and simple chalcogenides (e.g., SiTe [11]). OTS is the most mature selector technology that has been used for the commercial 3D X-point technology (to be discussed in Section 5.2). Another type of threshold I–V selector is leveraging the fast self-dissolution of metallic conductive filament (or the poor retention) in conductive bridge random access memory (CBRAM) [12, 13]. The potential problem for this CBRAM-based selector is the relatively slow turn-off speed and significant variability in threshold voltage. One general design challenge for the threshold I–V selector is that once the threshold switching occurs, most of the

TABLE 5.2
Survey of the Representative Selector Devices Reported in the Literature

Type	Stack	Voltage Range	Current Drivability	Nonlinearity	Endurance	Reference
Exponential I-V	Ni/TiO$_2$/Ni	−4 V to +4 V	0.1 MA/cm^2	10^3	>10^6	IEDM 2011
	Pt/TaO$_x$/TiO$_2$/TaO$_x$/Pt	−2.5 V to +2.5 V	32 MA/cm^2	10^4	>10^{10}	VLSI 2012
	TiN/a-Si/TiN	−3 V to +3 V	1 MA/cm^2	1.5×10^3	>10^6	IEDM 2014
	Ru/TaO$_x$/W	−4 V to +2.5 V	1 MA/cm^2	5×10^4	>10^{10}	IEDM 2016
	MIEC	−1.6 V to +1.6 V	50 MA/cm^2	10^4	N/A	IEDM 2012
Threshold I-V	TiN/NbO$_2$/W (IMT)	0.9 V–1 V	10 MA/cm^2	50	N/A	IEDM 2015
	TeAsGeSiSe based OTS	1.5 V–2 V	11 MA/cm^2	10^3	10^8	IEDM 2012
	SiTe based OTS	0.6 V–0.9 V	10 MA/cm^2	10^6	5×10^5	VLSI 2016
	FAST (CBRAM)	0.1 V–0.9 V	5 MA/cm^2	10^7	10^8	IEDM 2014
	Cu/doped-HfO$_2$/Pt (CBRAM)	0.05 V–0.4 V	4.1 MA/cm^2	10^7	10^{10}	IEDM 2015

Note: For exponential I-V type, the voltage range is the max voltage where the current density is measured; for the threshold I-V type, the voltage range is the hold voltage and the threshold voltage.

Emerging Non-volatile Memories

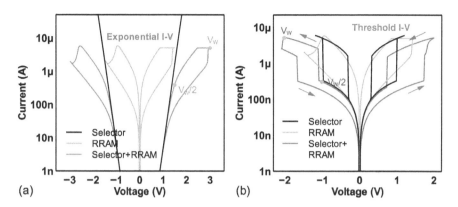

FIGURE 5.15 Representative I–V characteristics for (a) a bipolar RRAM with exponential I–V selector and (b) a bipolar RRAM with threshold I–V selector.

applied voltage will transfer from the selector to the memory cell, therefore, there is a risk of read disturb. The parameters of the selector (e.g., threshold voltage and hold voltage) need to be co-designed carefully with the memory device characteristics.

5.2 PHASE CHANGE MEMORY (PCM)

5.2.1 PCM Device Physics

PCM relies on the electrically triggered phase transition in the chalcogenides between the crystalline state and the amorphous state. Note that chalcogen is the element in group XI (except oxygen) in the periodic table. Typical PCM material is based on the Ge-Sb-Te (GST) alloy (e.g., $Ge_2Sb_2Te_5$). The crystalline phase has ordered long-range crystal structure and thus has lower free energy, corresponding to the LRS. On the other hand, the amorphous phase has a random and disordered atomic configuration thus has higher free energy, corresponding to the HRS. The reversible transition between the crystalline phase and the amorphous phase is essentially modulated by the temperature profile within the GST material. Figure 5.16 shows the temperature

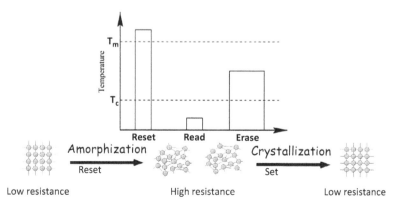

FIGURE 5.16 Principle of PCM and the temperature profile during the reset/read/set cycle.

profile during the reset/read/set cycle of the PCM. A short but high amplitude pulse that raises the temperature above the melting temperature (T_m) will melt the GST material. The atoms gain high kinetic energy and start moving around. If a quench process occurs by quickly removing the pulse, those atoms will not have sufficient time to settle, remaining in their disordered locations. This quench process will result in an amorphous phase after the reset. A very low amplitude read pulse will not change the amorphous phase. Then a long but medium amplitude set pulse will raise the temperature above the crystalline temperature (T_c) of the GST material. The atoms will start the thermally assisted vibration from the disordered locations, and they may gradually find their minimum energy locations as part of the recrystallization process if given sufficient time. The typical GST parameters T_m is 500 °C–700 °C, and T_c is 150 °C–300 °C, t_{reset} is ~10 ns, and t_{set} is ~100 ns. It should be noted that such temperature profile could be generated optically, for example, by laser annealing, which becomes the principle of optical disks such as CD/DVD that use similar GST material. Here in this chapter, PCM refers to the electrical memory where the temperature profile is generated by the current-induced Joule heating. The GST thin film is primarily fabricated by physical vapor deposition (PVD) methods, such as sputtering, and with a recent demonstration by atomic layer deposition (ALD).

Figure 5.17 shows the typical electrical characteristics of the PCM. Figure 5.17(a) shows the quasi-DC I–V sweep using the current forcing-voltage sensing method. If the device starts from HRS with an amorphous phase, a snapback occurs with a threshold voltage (V_{th}) following a set transition. If the device starts from LRS with crystalline phase, the current sweep will not trigger the reset transition, as the reset requires a fast quench process that is not available by quasi-DC sweep. Figure 5.17(b) shows the resistance evolution under current pulses (e.g., reset pulse 10 ns, and set pulse 100 ns). If the device starts from LRS, with a sufficiently high amplitude pulse (e.g., 400 µA), the melting process occurs and the resistance increases substantially. If the device starts from HRS, as the current pulse amplitude increases (from 150 µA to 300 µA), the amorphous phase undergoes the crystallization process and the resistance gradually decreases. However, if the current amplitude becomes too high into the reset regime, a similar quench process will occur. It should be noted that PCM is operated

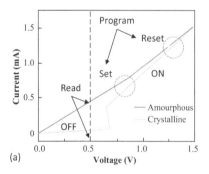
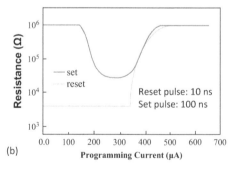

FIGURE 5.17 Typical electrical characteristics of the PCM: (a) quasi-DC sweep mode; (b) pulsing mode.

by the unipolar pulses (i.e., the same polarity of the voltage). Therefore, designing a precise waveform is necessary for correct set and reset operations in PCM.

There are two common types of PCM device structures, as shown in Figure 5.18(a). The first one is the mushroom cell, where GST is deposited on the planar surface and the active device switching region is defined by the underlying heater via (e.g., W or TiN plug). The second one is the confined cell, where the GST is filled into the trench where the active switching region is surrounded by the insulator with low thermal conductivity. The mushroom cell is easy to be fabricated, but the confined cell is preferred for a more efficient write operation. Since PCM relies on Joule heating to increase the temperature above T_m for quench, the reset current becomes the limiting factor for the power consumption. As shown in Figure 5.18(b), the reset current is a strong function of the active device area, as the volume for the GST quench varies with the device area. The confined cell requires approximately half of the reset current than the mushroom cell at the same equivalent cell diameter. This is because the confined cell has a geometry that the generated heat is more concentrated and difficult to dissipate, and thus it is easier to reach the T_m. Nevertheless, the required reset current of the confined PCM cell is still demanding. Roughly speaking, at 20 nm cell size, the reset current is still around 100 µA. The projected trend shows that to reach a desirable 10 µA level that the minimum-sized selection transistor could easily deliver, the PCM cell size has to scale toward the sub-5 nm regime.

PCM is capable of multilevel cell (MLC) operation, as shown in Figure 5.19. In principle, both set and reset processes could be exploited for switching into intermediate states. The gradual set operation is performed by adjusting the set pulse amplitude or the ramp-down slope of the set pulse. The partial crystallization is realized by percolation paths formed in the amorphous region. On the other hand, the gradual reset operation is performed by adjusting the reset pulse amplitude. The higher the amplitude of the reset pulse, the more is the heat generated; thus, a larger volume of the amorphous region will be obtained after the melting-quench process. In practice, a gradual set is easier to manipulate, but the feasibility of a gradual reset depends on the cell structure. For example, the amorphous volume of a mushroom cell is easier

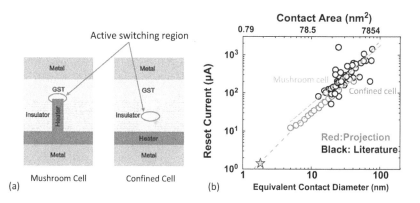

FIGURE 5.18 (a) Common types of the PCM device structures: mushroom cell and confined cell. (b) PCM reset current as a function of the equivalent contact diameter of the cell.

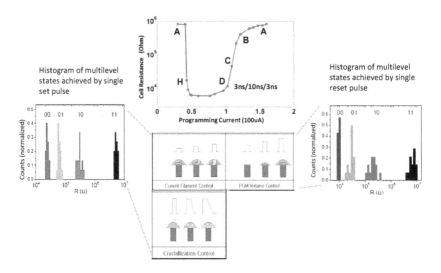

FIGURE 5.19 Principle of MLC operation in PCM, programming waveform and resulting resistance distribution of a 2 bits per cell.

to be adjusted by the reset pulse amplitude as the heat is spreading out, while the confined cell is more difficult to achieve multilevel states in reset as the volume is already confined and it may be melted as a whole. Figure 5.19 also shows the example of the measured resistance distribution of 4-level (2 bits per cell) in the PCM array with a single set/reset pulse. It is usually necessary to use the write-and-verify scheme to further tighten the distribution (similar to the ISPP used in NAND Flash as discussed in Section 4.4).

5.2.2 Reliability of PCM

There are two mechanisms affecting the data retention of PCM. The first one is the resistance drift. After the write operation, PCM cell resistance tends to drift over time toward higher resistance in a short-term timescale (e.g., from μs to hours). As shown in Figure 5.20(a), LRS, HRS, and intermediate states all exhibit the drift behavior. The higher the resistance, the severer the drift is. It is apparent that drift is a key challenge for the MLC operation, as the resistance may cross over the references between the intermediate states over time. The drift in PCM is believed to be associated with the structural relaxation of the amorphous phase, which is essentially a non-equilibrium and meta-stable state resulting from a rapidly quenched process. A following structural relaxation might also involve the annealing out of defects, resulting in lower trap density. Since the current conduction is dominated by trap-assisted tunneling, lower trap density leads to lower conductance and higher resistance. The resistance drift could be empirically fitted by the equation below:

$$R = R_0 \left(t/t_0\right)^v \quad (5.1)$$

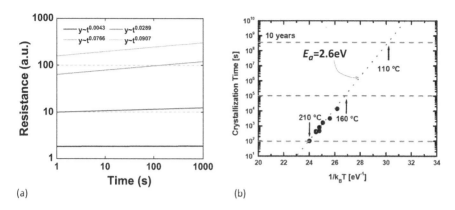

FIGURE 5.20 (a) Drift behavior of PCM; (b) Arrhenius plot for a PCM cell retention.

where R_0 is the initial resistance, t is the time, t_0 is the time normalization factor, ν is the drift coefficient.

The second mechanism is elevated temperature-induced crystallization, which is typically a long-term effect (e.g., from months to years). At an elevated temperature, the amorphous phase may spontaneously crystallize itself given sufficient thermal kinetic energy. As a result, HRS may gradually switch to LRS. For any thermal activation process, the Arrhenius law could be applied:

$$t = t_0 \exp(-E_a/kT) \quad (5.2)$$

where t is the time to crystallization, t_0 is the time coefficient, E_a is the activation energy, and kT is the thermal energy. Figure 5.20(b) shows the Arrhenius plot (log(t) versus 1/kT) for a PCM cell. High-temperature baking experiments could be performed to accelerate the crystallization (e.g., in the temperature range of 160 °C–210 °C and record the time when a predefined LRS level is reached). Then E_a could be extracted and data retention in the timescale that is of interest (e.g., 10-year lifetime) could be extrapolated. If using the electron volt (eV) as the unit for kT, the slope of this fitted line is E_a (2.6 eV in this case), as given by the equation below,[4]

$$E_a = \frac{2.3 \log_{10}(t_2/t_1)}{\left(\frac{1}{kT_2} - \frac{1}{kT_1}\right)} \quad (5.3)$$

(T_2, t_2) and (T_1, t_1) are two x-y-coordinates on the fitted line of the Arrhenius plot. It is noted that 10 years equals ~3 × 10^8 s; therefore, in this example, the PCM cell could maintain its HRS state at 110 °C for 10 years.

Due to the melting-quench nature of PCM, thermal cross-talk is a potential problem. Thermal cross-talk refers to the resistance disturbance of the victim cells that are adjacent to the selected cells being written, as shown in Figure 5.21(a). Figure 5.21(b) shows the simulated temperature profile as a function of the distance to the selected cell. The selected cell needs to reach the melting temperature (e.g., 700 °C)

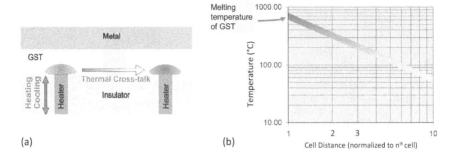

FIGURE 5.21 (a) Thermal cross-talk problem between adjacent cells of PCM. (b) Simulated temperature profile as a function of the distance to the selected cell.

for the reset, and the neighboring cell may face relatively high temperatures (e.g., 250 °C). Then the adjacent cell (if in HRS) undergoes the crystallization process. According to the data retention plot (e.g., in Figure 5.20(b)), the crystallization time is ~1 s at 250 °C. This means that if the selected cell is continuously being reset for millions of cycles, then the resistance of the victim cells may substantially change. The thermal cross-talk may become a more server problem with PCM's downscaling. As the cells become closer in distance, the temperature profile may spread out over more neighboring cells.

Cycling endurance determines how many times that the PCM can be written. The two main endurance failure modes of a PCM cell are either stuck at HRS or stuck at LRS. Stuck at HRS is mainly caused by the void formation at the bottom electrode interface after switching cycles. The other failure mode, stuck at LRS, is often caused by the elemental segregation upon repeated cycling. For example, Sb enrichment at the bottom electrode, which could have a lower crystallization temperature, leads to easier crystallization of the active region. Figure 5.22 shows a representative resistance evolution over cycling for both types of endurance failure. Different failure modes may occur on different cells in one array. Typical PCM could sustain 10^6–10^9 cycles, but it is difficult to reach 10^{12} cycles.

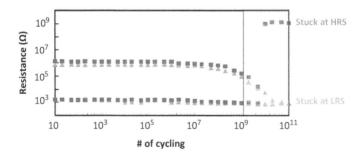

FIGURE 5.22 Representative resistance evolution over cycling for both types of endurance failure in PCM.

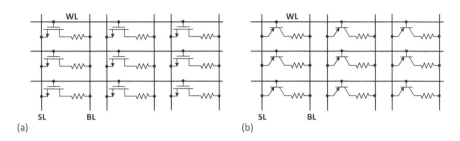

FIGURE 5.23 Two types of 1T1R array for the PCM integration: (a) with MOSFET as selection transistor; (b) with BJT as selection transistor.

5.2.3 ARRAY INTEGRATION OF PCM

When PCM cells are integrated into arrays, the 1T1R array could be used as shown in Figure 5.23. Here, the selection transistor could be the regular MOSFET. However, as aforementioned discussed, PCM's reset current is typically higher than the minimum-sized MOSFET could offer. For example, at the 20 nm node, the PCM reset current could be >100 μA; thus, at least W/L = 3 is required for a regular MOSFET to deliver such drive current. On the other hand, the bipolar junction transistor (BJT) could provide a higher current density at the same width as the MOSFET counterpart. Therefore, the BJT is sometimes used as the selection transistor (as demonstrated at 28 nm node by STMicroelectronics [14]), though the BJT's availability at the advanced technology node is rather limited. Table 5.3 shows the survey of state-of-the-art PCM prototype chips (as of 2020). These 1T1R-based PCM chips are primarily targeted for embedded Flash replacement [15, 16] for on-chip code storage (e.g., for microcontrollers).

TABLE 5.3
The Survey of the Recent PCM Prototype Chips

Tech. Node	STMicroelectronics 28 nm	STMicroelectronics 28 nm	TSMC 40 nm
	IEDM 2018	IEDM 2020	IEDM 2019
Target app.	eFLASH	eFLASH	eFLASH
Bit-cell structure	1T-1R (5V I/O MOSFET)	1BTJ-1R (5V I/O BJT)	1T-1R
Bit-cell size	0.036 μm²/45.9 F²	0.019 μm²/24.2 F²	N/A
R_{ON}/R_{OFF} (Ω)	14.39 K/748.89 K	29.1 K/284.36 K	4.34 K/1.22 M
On-off ratio	52	~9.8	~281
Write voltage or current	200 ~ 300 μA	~300 μA	~300 μA (reset)
Write pulse width (ns)	N/A	N/A	100 ns (set)
Read pulse width and current	5–40 μA	5–25 μA	N/A
Write endurance	> 10⁶	> 10⁷	> 2×10⁵
Retention	> 10 years @ 150°C	N/A	> 10 years @ 120°C

5.2.4 3D X-POINT

To achieve higher integration density for standalone memory applications, a cross-point array (assisted with selector) is preferred. Early demonstration of the cross-point array with PCM relies on the epitaxial silicon diode as the selector. Silicon diode with rectifying I–V characteristic and high nonlinearity is compatible with the PCM, as PCM could be operated in a unipolar manner with the same voltage polarity (e.g., only positive bias but varying the set/reset pulse width). In 2012, a 1 Gb prototype chip with a 20 nm half-pitch cross-point array was reported by Samsung with an integration of PCM and silicon diode, achieving 40 MB/s internal write bandwidth, and 120 ns read cycle time [17]. Even though such performance of the 2D cross-point array is superior to the 2D NAND Flash at that time, the integration density is not competitive yet as 3D NAND soon took over the market in the mid-2010s.

Therefore, the standalone PCM is re-targeted at the storage class memory as a layer between DRAM and NAND Flash. As DRAM continued the scaling into 1z node in the late 2010s, PCM needs to be 3D stackable to achieve higher integration density than DRAM. In such a context, a 3D X-point array becomes a must. However, the epitaxial silicon diode is not feasible for 3D integration as it requires a single crystalline silicon substrate to start with. Simply depositing amorphous or polycrystalline silicon at BEOL will have a large leakage current due to the defects; thus, the I–V nonlinearity is not sufficient for cross-point memory. Hence, developing a new type of selector that is 3D stackable is necessary. To best match the PCM characteristics, an OTS selector is usually adopted [18]. OTS has similar material components as PCM such as chalcogenides as discussed earlier in Section 5.1. OTS exhibits the threshold switching where the off-state abruptly turns on above the threshold voltage. Unlike PCM, OTS does not undergo structural changes (i.e., the crystallization process). One possible reason could be that OTS has a relatively high crystallization temperature; thus, the Joule heating is not sufficient to trigger the permanent transition. As a result, only electronic switching occurs with a large current flow. When the voltage is removed or lower than the hold voltage, OTS switches off spontaneously.

In 2017, Intel and Micron jointly announced the commercialization of 3D X-point technology based on PCM and OTS, resulting in Optane and X-100 high-end solid-state drive (SSD) products, respectively. As shown in Figure 5.24, the first-generation design is a 2-layer integrated cross-point array with 20 nm half pitch, where the peripheral CMOS circuits are hidden underneath the array. A reverse engineering report (by Tech Insights [19]) revealed that the PCM is based on GST material, and the OTS is based on Ge-Si-Se-As alloy. The commercial chip has a capacity of 128 Gb, a write bandwidth of 35 GB/s, and a read cycle time of 100 ns, making it competitive to be storage class memory between the DRAM and the NAND Flash. A more intuitive metric is the integration density in terms of Gb/mm², the first-generation 3D X-point is 0.62 Gb/mm², while contemporary DRAM products (e.g., in 1x nm node) is about 0.19 Gb/mm² and contemporary 3D NAND product (e.g., 64-layer with TLC) is 5.6 Gb/mm². In 2020, Micron announced a cease of further 3D X-point technology development, while Intel reported a second-generation of 3D X-point which has a 4-layer cross-point array.

Emerging Non-volatile Memories

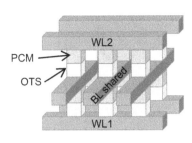

1st Generation 3D X-point Attributes (Intel/Micron)	
Attribute	Value
Cell Feature Size (half pitch)	20 nm
# Decks	2
Die Capacity	128 Gbits
Bank Access Size	16 Bytes
# Independent Banks	16
Read Latency per Bank	100 ns
Write Latency per Bank	500 ns
Write Bandwidth (GB/s)	~ 35 GB/s

FIGURE 5.24 3D X-point array and the key metrics of its first-generation chip.

5.3 RESISTIVE RANDOM ACCESS MEMORY (RRAM)

5.3.1 RRAM Device Physics

RRAM relies on the conductive filamentary formation and rupture mechanism in an insulator thin film between two electrodes and thus is capable of reversible switching between the insulating state and the conducting state. RRAM can be classified into two major classes: (1) oxide RRAM (OxRAM) where the conductive filament consists of oxygen vacancies; (2) conductive bridge RAM (CBRAM) where the conductive filament is made of metallic atoms. Figure 5.25(a) shows the switching mechanism of RRAM. In OxRAM, the oxygen vacancies are created by a soft breakdown under sufficient applied electric field, the oxygen atoms are ionized (i.e., oxygen ions) and migrated toward the top electrode interface (temporarily stored in a reservoir at one metallic capping layer), and thus a conductive filament is formed. Under the reverse electric field, oxygen ions can migrate back to annihilate oxygen vacancies, thus rupturing the filament. In CBRAM, the metallic atoms are ionized from one of the active metal electrodes (e.g., Ag or Cu, or their alloys) by the applied electric field, and migrated into the insulator layer that is typically chalcogenide or oxide, resulting in a metallic filament. Similarly, under the reverse electric field, the ionized metallic atoms could migrate back toward the top interface, thus rupturing the filament. In the following, OxRAM is used as the default type of RRAM for discussions because OxRAM is more popular in industrial demonstrations. Within OxRAM, HfO_x, and TaO_x have become the dominant oxide materials for prototyping and commercialization.[5] ALD is the common fabrication method for oxide thin film deposition.

It should be noted that many fresh RRAM cells require an initial forming process with a relatively high voltage (e.g., 3–5 V) to initialize the filament and trigger the subsequent switching at a lower voltage (e.g., 1–3 V). After forming, RRAM generally exhibits the bipolar switching I–V characteristics as shown in Figure 5.25(b). The set process is usually abrupt while the reset process is usually gradual. To achieve multilevel cell (MLC) operation, there are two ways: (1) use different set compliance currents (e.g., modulated by the selection transistor's gate voltage) to adjust the

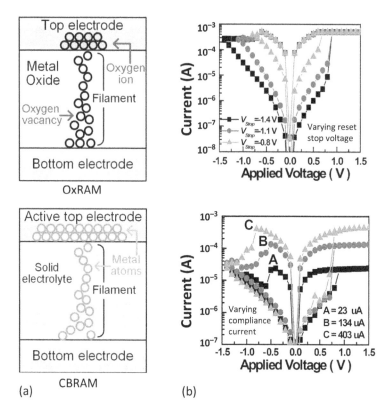

FIGURE 5.25 (a) RRAM switching mechanism in OxRAM and CBRAM; (b) representative I–V characteristics in RRAM showing the MLC achieved by varying either reset voltage or set compliance current.

diameter or strength of the filament; (2) use different reset voltages to weaken the filament or modulate the tunneling gap distance between the residual filament tip and the electrode.

Since the RRAM switching has an exponential relationship between the switching speed and the applied voltage. Figure 5.26 shows the required set/reset voltage amplitude to switch the RRAM cell at different pulse widths for the HfO_x RRAM with 10 nm cell size [20]. The reset speed has a stronger dependence on voltage than the set speed. A reset speed of ~10 ns is feasible with a large voltage (i.e., 2.5 V), and a reset speed of ~1 μs is needed if the voltage is lowered (i.e., to 2 V). Hence, a design trade-off exists between the write speed and the write voltage.

5.3.2 Reliability of RRAM

RRAM has a typical cycling endurance in the range of 10^6–10^9. Figure 5.27 shows an example of cycling endurance characteristics of HfO_x RRAM [21]. It is noted that the endurance failure modes depend on the programming conditions. If the set condition is stronger (e.g., with higher compliance current by larger WL voltage), the cell may

Emerging Non-volatile Memories 157

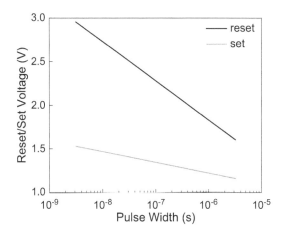

FIGURE 5.26 The relationship between set/reset voltage amplitude and pulse width for the HfO_x RRAM.

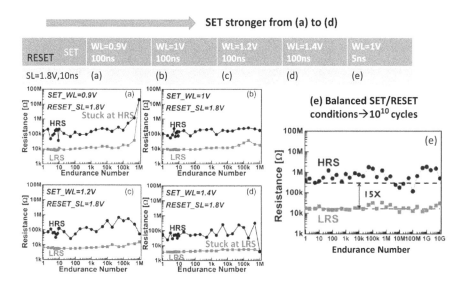

FIGURE 5.27 Example of cycling endurance characteristics of the HfO_x RRAM under different programming conditions.

be stuck at LRS after cycling. This is because excessive oxygen vacancies are formed during the set. On the other hand, if the reset condition is stronger, the cell may be stuck at HRS as the filament may be completely ruptured. With a balanced set/reset conditions, the cell may reach 10^{10} cycles. However, due to the cell-to-cell variation, it is difficult to optimize the set/reset conditions that work well for an entire array.

The RRAM's data retention could be evaluated by a varying-temperature acceleration test, similar to that in the PCM. Figure 5.28(a) shows an example of retention characteristics of HfO_x RRAM for both HRS and LRS at 150 °C, 200 °C, and 250 °C [22].

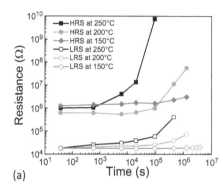 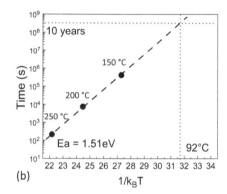

FIGURE 5.28 (a) Example of retention characteristics of the HfO$_x$ RRAM for both HRS and LRS. (b) The Arrhenius plot of RRAM retention.

In all cases, the cell resistance increases after baking due to the spontaneous dissolution of the filament. The higher the temperature, the faster the degradation is. Given a criterion of reference resistance between LRS and HRS, the time to retention failure could be extracted as a function of temperature, and the Arrhenius plot could be drawn as Figure 5.28(b). The extracted slope for activation energy E_a is 1.51 eV in this case, and the extrapolated 10-year lifetime could be maintained at a temperature of 92 °C.

5.3.3 Array Integration of RRAM

In general, RRAM shows attractive features such as low programming voltage (<3 V), fast switching speed (<100 ns), large resistance on/off ratio (>10×), reasonable endurance (>10^6 cycles), and better data retention (>years at 85 °C). In the early stage of RRAM development, 180 nm or 90 nm CMOS platforms were primarily used. For instance, in 2011, ITRI reported a 4 Mb 1T1R RRAM macro at 180 nm node featuring a small cell area (9.5 F^2), and 7.2 ns read/write random access for single-level-cell operation, and 160 ns for multi-level-cell (MLC) operation [23]. In 2012, Panasonic reported an 8 Mb two-layer cross-point RRAM macro with tunneling selector at 180 nm node featuring small equivalent bit area (2 F^2), 443 MB/s write throughput (64-bit parallel write per 17.2 ns cycle) and 25 ns read access [24]. In 2017, Winbond reported a 512 kb 1T1R RRAM at 90 nm node featuring a relatively small cell area (30 F^2), 100 ns read/write random access, and 10-year retention at 150 °C after >10^5 cycles [25]. Recent developments of RRAMs aim at low-cost embedded NVM solution at 28 nm node or 22 nm node that is more competitive than eFlash. Table 5.4 shows the survey of state-of-the-art RRAM prototypes (as of 2020). TSMC is providing RRAM solutions at 40 nm [26], 28 nm [27], and 22 nm [28], and Intel is offering RRAM solution at 22FFL platform [29].

In the early 2010s, standalone RRAM was once positioned for the storage class memory application. In 2013, SanDisk/Toshiba reported a 32 Gb 2-layer cross-point RRAM macro at a 24 nm technology node that has the largest capacity to date [30]. In 2014, Micron/Sony reported a 16 Gb 1T1R RRAM macro at 27 nm technology node featuring a small cell area (6 F^2), 200 MB/s write bandwidth, and a 1 GB/s read

TABLE 5.4
The Survey of the Recent RRAM Prototype Chips

Tech. Node	TSMC 22 nm	TSMC 28 nm	TSMC 40 nm	Intel 22 nm	Winbond 90 nm
	VLSI 2020	VLSI 2020	ISSCC 2018	ISSCC 2019	IEDM 2017
Target app.	eFLASH	eFLASH	eFLASH	eFLASH	eFLASH
Bit-cell structure	1T-1R	1T-1R	1T-1R	1T-1R	1T-1R
Bit-cell size	53 F^2	N/A	53 F^2	0.0484 μm^2/100 F^2	0.5 μm^2/31 F^2
R_{ON}/R_{OFF} (Ω)	N/A	N/A	Est. R_{ON} ~ 4 K	3–7 K/30 K	6–7 K/>500 K
On-off ratio	~4	~3	5–6	4–10	~100
Write voltage or current	1.62 V–3.63 V	N/A	1.4 V–2.4 V	N/A	2–4 V
Write pulse width (ns)	N/A	N/A	<1 μs	<10 μs	100–200 ns
Read pulse width and speed	10 ns/0.7 V	20 ns/0.2 V	9 ns/0.26 V	5 ns/0.7 V	10 ns/0.5 V
Write endurance	> 10^4	> 10^5	> 10^3	N/A	>2x10^5
Retention	N/A	N/A	N/A	N/A	>100 years @ 85 °C

bandwidth [31]. However, the industry soon realized that due to significant variations, large capacity RRAM is not viable for the commercial grade. The limited endurance of RRAM (even less than PCM) also limits its feasibility as storage class memory. Therefore, the notable endeavor to pursue standalone RRAM for the storage class memory ceased after the PCM-based 3D X-point technology was announced.

5.4 MAGNETIC RANDOM ACCESS MEMORY (MRAM)

5.4.1 MTJ Device Physics

The core element of MRAM is the magnetic tunnel junction (MTJ), which consists of two magnetic layers separated by a thin tunnel oxide barrier. The typical magnetic layer material is CoFeB or its variants, and the typical tunnel oxide material is MgO. One of the magnetic layers is designed to have a pinned magnetization orientation, usually assisted by the synthetic antiferromagnetic (SAF) structure that is immune to the external magnetic field. On the other hand, the other magnetic layer is able to switch its magnetization orientation depending on the external magnetic field. Figure 5.29 shows the hysteresis of MTJ resistance under an external magnetic field. When the two magnetic layers are in the parallel (P) state or the antiparallel (AP) state, the MTJ resistance is low or high, respectively. The resistance difference is caused by the spin-dependent quantum tunneling effect. As shown in Figure 5.30, when the two magnetic layers are in the AP state, the spin-polarized

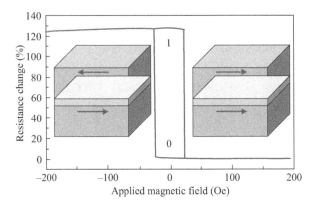

FIGURE 5.29 The hysteresis of MTJ resistance under an external magnetic field.

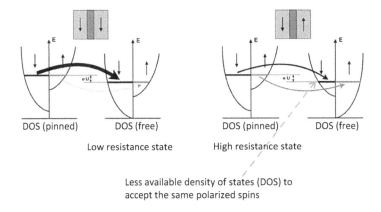

FIGURE 5.30 Diagram of energy vs. density of states (DOS) to illustrate the spin-dependent quantum tunneling effect of the MTJ in parallel state and anti-parallel state.

current will be more difficult to tunnel through the tunneling barrier. This is because the density of states in the spin-down state on the right-hand side is less populated to accept the spin-down polarized electrons from the left-hand side. The tunneling magneto-resistance ratio (TMR) is usually used to define the on/off ratio of the MTJ resistance, it is given by

$$\text{TMR} = (R_{ap} - R_p)/R_p \qquad (5.4)$$

where R_{ap} is the resistance of the antiparallel state, and R_p is the resistance of the parallel state. Generally, MTJ's TMR is in the range of 50%–200%. Here, TMR = 100% means on/off ratio = 2. It is noted that MRAM's on/off ratio is much lower than that of other NVM devices, thus demanding a more complex sense amplifier design to tolerate the tight margin. Due to the same reason, MRAM generally could not offer MLC capability.

Emerging Non-volatile Memories

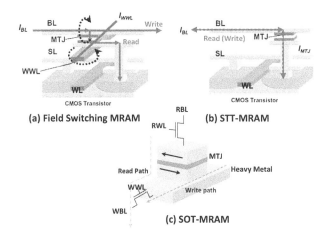

FIGURE 5.31 MRAM bit-cell options: (a) Field switching MRAM; (b) STT-MRAM; (c) SOT-MRAM.

With MTJ, there are a few device options for the MRAM integration, as shown in Figure 5.31. The first generation is the field switching MRAM, the second generation is the spin-transfer-torque (STT) MRAM, and the third generation is the spin-orbit-torque (SOT) MRAM. These three types of MRAM will be discussed in detail in the following sub-sections.

5.4.2 Field Switching MRAM

Since MTJ could be switched by the magnetic field, it is intuitive to design a device structure as shown in Figure 5.32(a) to have a write WL (WWL) underneath the MTJ (and it is separated by an isolation layer). The write operation is done by passing sufficient current to the WWL and the BL simultaneously. As shown in Figure 5.32(b), the magnetic field surrounding the wires is generated following the right-hand rule in electromagnetics. The selected cell is the one that is at the intersection of the selected WWL and BL, while the half-selected cells are along the WWL or along the BL. A similar half-select problem as discussed in the cross-point array exists here. To minimize potential write disturb, a toggle MRAM concept is adopted, which intentionally designs the elliptical anisotropy of the in-plane MTJ with its hard-axis along 45° between the WWL and the BL. Using a two-step pulsing scheme as shown in Figure 5.32(b), the magnetization orientation could be flipped. The read operation is done through the 1T1R structure, where the selection transistor is controlled by the read WL (that controls the gate of the transistor as shown in Figure 5.31(a)), and the current depending on the resistance state is sensed at BL.

The industrial development of the field switching MRAM started in the early 2000s. In the mid-2000s, 16 Mb field switching MRAM was demonstrated on IBM's 180 nm platform, where MTJ was integrated underneath M3 interconnect [32]. The equivalent cell size is 44 F^2. The write/read latency is 30 ns. At such speed, the write current on write WL and BL is 5 mA per cell, and the read current

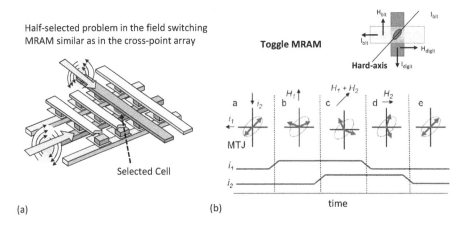

FIGURE 5.32 (a) Write mechanism of field switching MRAM; (b) timing and waveform design for toggle MRAM.

on BL is 1.56 mA per cell. Obviously, the ultra-high current requirement is the major challenge for scaling the field switching MRAM to more advanced technology nodes. The field switching MRAM did not gain much attention in the industry after the mid-2000s.

5.4.3 STT-MRAM

The interest in developing MRAM resurged after the discovery of the STT mechanism to switch the MTJ in the late-2000s. As shown in Figure 5.31(b), the STT-MRAM contains a 1T1R cell, where the write and read share the same path between BL and SL, and WL controls the selection transistor. Figure 5.33 shows the principle of the STT switching driven by the bipolar current between BL and SL. For the AP to P switching, a positive write voltage is applied to BL. When a sufficiently large

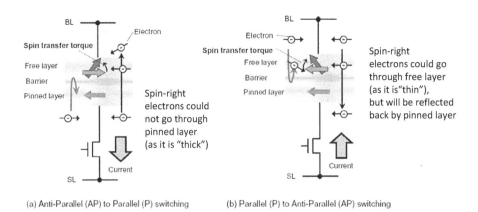

FIGURE 5.33 Principle of STT-MRAM switching mechanism.

current is flowing from the free layer to the pinned layer, the electrons are injected the other way. In this example, the magnetization of the pinned layer is pointing to the left; thus, the injected spin-right electrons could not travel through the pinned layer (and they are reflected backward), while only the spin-left electrons could tunnel through the oxide barrier. Then the spin-right electrons that are tunneled through will transfer their spin momentum to the free layer. The free layer's magnetization (originally pointing to the right) feels a torque opposite to the damping force. The precessional motion of magnetization is amplified and rotates to the left direction, which could be assisted by the thermal vibration. On the other hand, for the P to AP switching, a positive write voltage is applied to SL. When a sufficiently large current is flowing from the pinned layer to the free layer, the electrons are injected the other way. In this case, the free layer is sufficiently thin and could not block the spin-polarized current; thus, both spin-left and spin-right electrons will tunnel through the oxide barrier; however, spin-right electrons could not further travel through the pinned layer that is pointing to the left. The reflected back spin-right electrons will transfer their spin momentum to the free layer. The free layer's magnetization (originally pointing to the left) feels a torque opposite to the damping force. The precessional motion of magnetization is amplified and rotates to the right direction, which could be assisted by the thermal vibration. It is noted that there is an asymmetry in the bipolar switching, and usually, the AP to P switching is easier thus requiring less write current than the P to AP switching.

The primary advantage of STT over field switching is that the write current downscales with the device's active cell area. Given the same MTJ stack, the write current of the STT-MRAM is proportional to the cell area; thus, the write current density (J_c) becomes a more useful metric. The typical J_c is in the range of 1–4 MA/cm^2. If an advanced MTJ size is about 50 nm, then the write current is in the range of 25–100 µA, which is much lower than the mA-level requirement in the field switching MRAM. A selection transistor with W/L = 1–4 could offer such write current, which gives an estimated cell area around 16–64 F^2. With the scaling, the required write current scales proportional to F^2, which is a faster scaling factor than the transistor's drivability that scales proportional to F. As a result, the scaled STT-MRAM is able to use smaller width of the transistor even if the J_c is kept as a constant.

It should be pointed out that the write current density of STT-MRAM depends on the write pulse width. Figure 5.34(a) shows the typical relationship between the write current density (in the unit of MA/cm^2) and the write pulse width (ns). Typically, below 10 ns, the STT switching will enter the precession regime where the J_c increases rapidly with the decrease of the pulse width. Above 10 ns, the STT switching is assisted by the thermal activation, and most of the STT-MRAM tends to be operated in this regime. Due to the uncertainty of thermal vibration, the STT switching is also non-deterministic in this regime. The switching probability depends on the applied write voltage. Larger write voltage, larger switching current, and thus higher switching probability and fewer write errors. Figure 5.34(b) shows the write bit error rate (BER) of an STT-MRAM array as a function of write voltage. At a given voltage, a longer pulse will result in lower BER. In sum, optimized STT-MRAM is typically capable of a 10 ns write latency under 0.5 V write voltage.

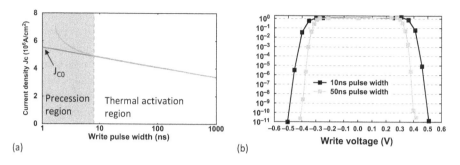

FIGURE 5.34 (a) The relationship between the write current density and the write pulse width in STT-MRAM. (b) The write bit error rate (BER) in STT-MRAM arrays as a function of write voltage.

STT-MRAM's data retention could be modeled by a two-state system (P and AP states) with an energy barrier in between, as shown in Figure 5.35(a). The retention failure is caused by the thermal vibration at elevated temperatures. Similar to PCM or RRAM, STT-MRAM's retention time could be drawn in the Arrhenius plot, as shown in Figure 5.35(b). The extracted slope is the activation energy (E_a). Thermal stability factor Δ is defined as E_a/kT. A wide range of Δ (30–80) has been reported in the literature for different MTJ stacks. Generally, the higher the Δ, the larger the J_c. This is because the more stable the cell is during the retention, the more difficult it could be flipped during the write. Therefore, a figure of merit J_c/Δ is often used to evaluate the quality of the MTJ stack, and a low J_c/Δ is preferred.

The cycling endurance of STT-MRAM is theoretically infinite, but it is practically limited by the tunnel oxide breakdown. As the spin-polarized current tunnel through the MgO barrier for many cycles, defects may be generated and accumulated till a breakdown occurs. In practice, a voltage stress accelerated test could be applied. The cycling endurance test is performed at a write voltage higher than the nominal

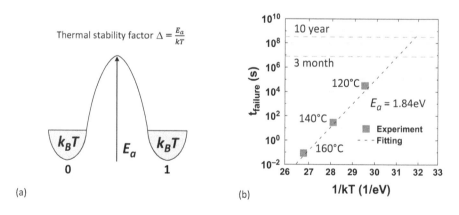

FIGURE 5.35 (a) A two-state model with an energy barrier in between two states (P "1" and AP/ "0"). (b) Arrhenius plot of STT-MRAM retention.

operation voltage, and the cycle-to-breakdown is recorded. Then, the endurance cycles at the nominal voltage are projected by extrapolation. For the oxide breakdown, typically there are two types of extrapolation model, one is the $1/E$ model that is often used in the higher field regime where the acceleration test is performed, and the other is the E model that is often more accurate in the lower field regime. Here, E is the electric field across the oxide. According to the $1/E$ model, the endurance cycles (c) is given by

$$c = c_0 \exp(E_0/E) \qquad (5.5)$$

According to the E model, the endurance is given by

$$c = c_0 \exp(-E/E_0) \qquad (5.6)$$

where c_0 and E_0 are fitting parameters. Figure 5.36 shows an example of the extrapolation of both methods. In either method, the projected endurance is more than 10^{14} cycles in the nominal write voltage (e.g., 0.5 V) even for the tail bits of the array (e.g., the 1 ppm line).[6] Compared to other NVMs, STT-MRAM has the best cycling endurance and has the potential to reach the endurance that is competitive to SRAM and DRAM.

What has been discussed so far is based on the in-plane MTJ. To enable further scaling toward more advanced nodes, the perpendicular MTJ is preferred. In-plane MTJ means that the magnetization orientation is in the same plane as the x-y plane of the thin film. Perpendicular MTJ means that the magnetization orientation is in the z-direction that is perpendicular to the x-y plane of the thin film. Figure 5.37 shows the comparison between in-plane MTJ and perpendicular MTJ. From the switching efficiency's perspective, perpendicular MTJ is more efficient because the thermal vibration direction is the same as the spin torque transfer path that could better assist the switching, while in-plane MTJ has different paths between these two. From the thermal stability's perspective, in-plane MTJ needs to maintain the x-y plane elliptical anisotropy; thus, a lateral elliptical shape that has a length larger than twice of width is generally required. On the other hand, perpendicular MTJ could use a

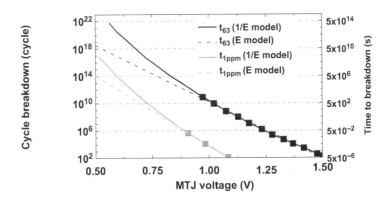

FIGURE 5.36 The extrapolation of STT-MRAM's cycling endurance using the E model and the $1/E$ model. The trends for 63% of the cells and the tail bit (1 ppm) are shown.

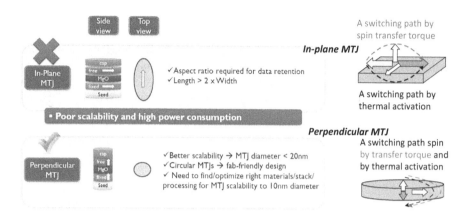

FIGURE 5.37 The comparison between in-plane MTJ and perpendicular MTJ.

minimum-size circular shape that offers better scalability toward sub-20 nm diameter. Considering both switching efficiency and thermal stability, perpendicular MTJ has a better figure of merit (J_c/Δ) than in-plane MTJ; thus, it has become the mainstream option for STT-MRAM technology today.

After almost a decade of industrial research and development, STT-MRAM has been commercially available by multiple vendors in the late 2010s. The array-level integration of STT-MRAM follows the 1T1R design principle discussed in Section 5.1, but different vendors may place the MTJ at different levels of the metal interconnect. There are different application scenarios for STT-MRAM. The first target is the standalone memory that aims to serve as the persistent memory (compatible with DRAM interface but being non-volatile), namely NVDIMM. In 2019, Everspin announced a 1 Gb STT-MRAM at 28 nm node with a DDR4 interface. It is noted that STT-MRAM-based NVDIMM has a significantly higher bit cost than the commodity DRAM, and the system-level benefits resulting from the non-volatility are yet to be demonstrated. Alternatively, many other vendors show interest in the embedded memory applications with STT-MRAM, either as the eFlash replacement at advanced node (28 nm or beyond) or as the last level cache. Table 5.5 shows the summary of STT-MRAM for eFlash replacement, with Globalfoundries offering at 22 nm node [33], Intel offering at 22 nm node [34], TSMC offering at 16 nm node [35], and Samsung offering at 28 nm node [36]. For eFlash replacement, data retention is of priority; thus, the thermal stability factor Δ needs to be high above 60, while it may trade-off with the lower endurance to around 10^6 cycles. Table 5.6 shows the summary of STT-MRAM for the last level cache replacement, with Globalfoundries offering at 22 nm node [37], IBM offering at 14 nm node [38], and Intel offering at 22 nm node [39]. For the last level cache replacement, cycling endurance is of priority with the target toward >10^{14}. As a result, the thermal stability factor Δ will be traded-off to around 20–40 and the data retention is lowered to seconds to minutes. Engineering the MTJ stack could enable different flavors that suit either the eFlash or the last level cache targets. For example, in Intel's FFL22 platform, simply reducing the MTJ cell size from 80 nm × 80 nm to 60 nm × 60 nm could switch the

TABLE 5.5
The Survey of the Recent STT-MRAM Prototype Chips for eFlash

Tech. Node	GF's 22 nm FDSOI	Intel 22 nm FFL	TSMC 16 nm FinFET	Samsung 28 nm
	IEDM 2020	IEDM 2018	IEDM 2020	IEDM 2019
R_{ON}/R_{OFF} (KΩ)	N/A	1.4/3.9	N/A	N/A
On-off ratio	2.48–3.12	2.8	N/A	2.8–3.2
MTJ size (nm²)	N/A	80 x 80	N/A	35^2–60^2
RA product (Ω·μm²)	N/A	9	N/A	N/A
Cell size (F²/μm²)	97.1/0.047	100 F²/0.0486 μm²	128.9 F²/0.033 μm²	45 F²/0.036 μm²
Write voltage (V)	1	N/A	N/A	1.05 V
Write pulse width (ns)	200 ns	20 ns–1 μs	50 ns	50 ns
Read pulse width (ns)	19 ns	10 ns	9 ns	40 ns
Write endurance	10^5–10^6	> 10^6	10^5	> 10^6
Retention	>20 years @150°C	10 years @ 200°C	N/A	10 years @ 105°C

TABLE 5.6
The Survey of the Recent STT-MRAM Prototype Chips for the Last level Cache

Tech. Node	GF 22 nm FDSOI	IBM 14 nm FinFET	Intel 22 nm FFL
	IEDM 2020	IEDM 2020	IEDM 2019
R_{ON}/R_{OFF} (KΩ)	N/A	7.87/19.21	2.5/7.0
On-off ratio	N/A	2.2–2.45	2.8
MTJ size (nm²)	<0.8x of eFLASH version	43^2 (35^2–60^2)	60 x 60
RA product (Ω·μm²)	N/A	14.55	9
Cell size (F²)	N/A	139.3/0.0273 μm²	100 F²/0.0486 μm²
Write voltage (V)	N/A	N/A	1.05–1.1
Write pulse width (ns)	10 ns	4 ns	20 ns
Read pulse width (ns)	<5 ns	N/A	4 ns/0.9 V or 8 ns/0.6 V
Write endurance	>10^{12}	>10^{10}	10^{12}
Retention	10s@125°C	1min @ 85°C	1s @ 110°C

properties from the eFlash to the last level cache. It is noted that STT-MRAM will be competing with RRAM and PCM for the eFlash replacement at 28 nm node or 22 nm node. However, STT-MRAM's fabrication process is more complex and thus the cost tends to be a disadvantage. Compared to other eNVM technologies, STT-MRAM exhibits one unique advantage: low write voltage (<1 V). Hence, STT-MRAM is compatible with the power supply voltage in an advanced logic process and thus holding the potential of scaling toward leading-edge nodes such as 7 nm or beyond. The compatibility and scalability toward the leading-edge node are essential for

STT-MRAM's competitiveness as the last level cache. This is because STT-MRAM cell density is around two times better than SRAM at the same technology node. Therefore, one generation behind the state-of-the-art SRAM is the minimum requirement for STT-MRAM to present the integration density benefits.

5.4.4 SOT-MRAM

Spin-orbit-torque (SOT) is a relatively new mechanism to switch the MTJ. As shown in Figure 5.31(c) earlier, an MTJ sits on top of the heavy-metal thin film, so the MTJ's free layer directly interfaces with the heavy-metal wire (such as Pt, W, or Ta). Spin–orbit interactions in a heavy metal can switch the magnetization orientation of the adjacent free layer when a sufficient write current is passing through the heavy-metal wire. Specifically speaking, the electron flow generates a spin current perpendicular to the heavy-metal thin film and a spin accumulation at the surface of the heavy metal due to the spin–orbit coupling effect. The accumulated spin momentum is transferred to the free layer's magnetization. Different from the field switching where the WWL wire is separated from the MTJ stack, SOT switching requires the heavy metal directly contacting the MTJ stack. For the SOT-MRAM bit cell, typically two selection transistors are required, one as the writing transistor connecting to the bottom heavy metal, and the other one as the reading transistor connecting to the top electrode of MTJ. Therefore, the read and write paths are decoupled. The write current flows through the heavy-metal wire but not through the MTJ's MgO tunnel barrier; so, the stress voltage across the MgO is much lower than that for STT switching, which offers the expectation of higher endurance cycles. Obviously, the drawback of SOT-MRAM is a larger cell area than STT-MRAM due to the three-terminal configuration and additional selection transistor. The primary advantage of SOT switching compared to STT switching is faster switching speed, and theoretically, sub-ns switching is achievable [40]. The improved endurance cycles and faster write latency make SOT-MRAM more compelling for the last level cache replacement. However, the write current density and the write energy per bit is still 10–100× higher than the SRAM counterparts [41], making it attractive only for the standby-frequent applications.

There are three types of SOT-induced magnetization switching, depending on the direction of the magnetization induced by the write current, as shown in Figure 5.38. Assuming write current flows the x-direction in the heavy metal, Type X/Y/Z means the easy-axis of the magnetization (M) of the free layer is in parallel with the x-direction/y-direction/z-direction, respectively. Type X and Type Y employ the in-plane MTJ, while Type Z employs the perpendicular MTJ. Type X and Type Z generally require an external magnetic field to break the symmetry, while Type Y does not need the external magnetic field, but it suffers from a slower switching speed. Structural engineering is possible to break the symmetry and enable external magnetic field-free switching. As shown in Figure 5.39, the canted SOT-MRAM cell [42] uses an angle to align the elliptical shape with respect to the current flow direction in heavy-metal wire, which achieved 0.35 ns switching without external magnetic field and still maintains a high thermal stability factor $\Delta = 70$.

Emerging Non-volatile Memories 169

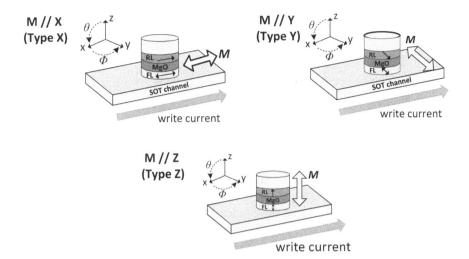

FIGURE 5.38 Three types of SOT-induced magnetization switching, depending on the direction of the magnetization induced by the write current. The trajectory of the spin switching is shown.

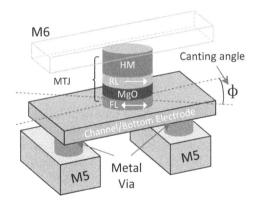

FIGURE 5.39 The canted SOT-MRAM that breaks the symmetry and enables external magnetic field-free switching.

5.5 FERROELECTRIC MEMORIES

5.5.1 Ferroelectrics Device Physics

Ferroelectricity is a phenomenon of spontaneous polarization in a special class of dielectrics. In normal dielectrics, the dipole moments are aligned up when the external electric field is applied but are randomized after the external electric field is removed. Instead, in ferroelectrics, the dipole moments are still aligned up after the external electric field is removed. Hence, the net polarization (P, i.e., the surface charges per unit area) of ferroelectrics is non-zero at zero bias, resulting in finite

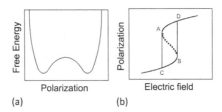

FIGURE 5.40 The Landau physical picture of ferroelectrics: (a) the free energy versus the electric field; (b) the polarization versus the electric field.

remnant polarization (P_r). Figure 5.40(a) shows the free energy (G) versus the electric field (E) in the ferroelectric material. Above the Curie Temperature, the material is dielectric, and below the Curie Temperature, the material is ferroelectric. The free energy of ferroelectric material could be described with a double valley separated by a barrier, following the Landau theory for the phase transition as follows:

$$G = \left(\frac{\alpha}{2}\right)P^2 + \left(\frac{\beta}{4}\right)P^4 + \left(\frac{\zeta}{6}\right)P^6 - EP \tag{5.7}$$

where α, β, and γ are fitting parameters. The two valleys correspond to the bi-stable states of the system (polarized up and polarized down). It is noted that the electric field is obtained when the derivative of the free energy with respect to the polarization is zero as follows:

$$\frac{\partial G}{\partial P} = 0 \rightarrow E = \alpha P + \beta P^3 + \zeta P^5 \tag{5.8}$$

Figure 5.40(b) shows the P–E hysteresis loop of the polarization versus the electric field. If the transition between A and B could be stabilized, there exists a negative "differential" capacitance regime [43]. But in normal voltage sweeps, B to D transition or A to C transition will be observable. With the external electric field applied, the barrier will be lowered accordingly with respect to the polarity of the field. Therefore, the transition from one state to the other will occur at the threshold electric field, namely the coercive field (E_c). The Landau physical picture applies well to single-domain ferroelectric switching.

The conventional ferroelectric materials are mostly perovskite oxides such as Lead Zirconate Titanate (PZT). However, the PZT fabrication process is incompatible with advanced silicon CMOS technology nodes. In the early 2010s, the discovery of ferroelectricity in HfO_2-based materials resurged the interests of CMOS compatible ferroelectrics. As is known, HfO_2 has been widely used by the industry in high-k/metal gate technology for advanced logic transistors. It should be noted that HfO_2 used in the high-k gate stack is usually in the amorphous phase (beneficial to minimize the threshold voltage variation caused by grain boundaries). If properly engineered (e.g., with high-temperature annealing), doped HfO_2 could show ferroelectricity in its crystalline or polycrystalline phases. Pure HfO_2 generally forms the non-polar monoclinic phase. Introducing dopant to the HfO_2 crystal structure could break up

Emerging Non-volatile Memories 171

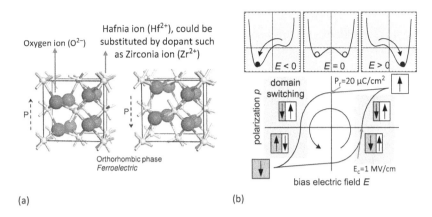

FIGURE 5.41 (a) Crystal structure of ferroelectric HfO$_2$ in the polar orthorhombic phase. (b) The P–E hysteresis loop showing gradual transition with multi-domain switching.

the symmetry and transform it into the polar orthorhombic phase. Depending on the position of the dopant, there could be a net charge or dipole pointing up or pointing down, as shown in Figure 5.41(a). It has been experimentally reported that dopant variants such as Si, Zr, Al, Y, Sr, Gd, La, etc., could induce ferroelectricity in HfO$_2$. Among all the reported HfO$_2$-based ferroelectrics, Si-doped HfO$_2$ and Zr-doped HfO$_2$ (or Hf$_x$Zr1-$_x$O$_2$ alloy, HZO) are becoming popular candidates for device integration. Si-doped HfO$_2$ requires a relatively high annealing temperature (700 °C–900 °C) for crystallization; thus, it is more suitable for the front-end-of-line (FEOL) integration or the gate-first process. On the other hand, HZO needs relatively low annealing temperatures (350 °C–450 °C) for crystallization; thus, it is more compatible with the BEOL integration or the gate last process.

The deposited thin film of doped HfO$_2$ may contain many microstructural grains, and the grains form the basis of the electrical domains. Due to the variations (e.g., E_c) between grains, the domain flipping may follow a nucleation process and some domains flip first in some locations, and then the flipping propagates through the entire area via domain-to-domain interactions. Owing to the multi-domain nature of the polycrystalline ferroelectric thin film, in practice, the P-E transition loop is not as abrupt as the Landau theory projects, as shown in Figure 5.41(b). The typical parameters of HfO$_2$-based ferroelectrics are shown. For instance, $P_r \sim 20$ μC/cm^2 and $E_c \sim 1$ MV/cm. The gradual transition of P–E hysteresis could be empirically described by the generalized Preisach model [44] featured with a tanh function, as formulated below:

$$P = P_s \tanh\left(s \cdot (E - E_c)\right) + P_{\text{offset}} \quad (5.9)$$

$$s = \frac{1}{E_{c+} - E_{c-}} \log\left(\frac{P_s + P_r}{P_s - P_r}\right) \quad (5.10)$$

$$E_c = \begin{cases} E_{c+}, \text{if } E > 0 \\ E_{c-}, \text{if } E < 0 \end{cases} \quad (5.11)$$

where P_s is the saturation polarization (i.e., the maximum polarization), P_r is the remnant polarization, s is the slope parameter of the P–E hysteresis loop, E_{FE} is the electric field in the ferroelectrics, P_{offset} accounts for the shift of hysteresis loop center along the polarization axis, and E_{c+} and E_{c-} are the forward and backward sweep coercive fields of the ferroelectrics that dictate the shift of hysteresis loop center along the electric field axis, respectively.

To experimentally extract the ferroelectric parameters such as E_c, P_r, and P_s from the P–E loop, a positive-up-negative-down (PUND) testing protocol is widely used on the ferroelectric capacitor, where the ferroelectric material is sandwiched between two electrodes. The PUND protocol aims to distinguish the ferroelectric polarization switching from the normal dielectric charging/discharging effects. Figure 5.42(a) shows the applied voltage waveform of PUND measurement, during which the real-time transient current is monitored. The first positive pulse flips the domains up, resulting in a relatively large current including both polarization switching current and displacement current. If the device is fully polarized, the second positive pulse only induces a relatively small displacement current. By extracting the second current from the first current, the actual polarization switching current is obtained for the positive sweep. A similar principle applies to the negative sweep. By integrating the current and the time during the test, the charges can be calculated, and the entire P–E loop is thus reconstructed as shown in Figure 5.42(b). Hence, the switching polarization (P_{sw}, also equal to $2P_r$) could be extracted from the PUND test.

To explore the intrinsic reliability effects, a simple ferroelectric capacitor is usually employed. The cycling endurance of ferroelectrics generally follows a wake-up and then a fatigue process. Figure 5.43 shows the endurance testing protocol and an example of the P_{sw} measured from an HZO-based ferroelectric capacitor. A positive/negative rectangle pulse sequence is applied, and at each checking point, a PUND test is performed to extract the P_{sw}. It is noted that the polarization switching window is initially low in the fresh cell. As the cell experiences more cycling, the P_{sw} increases, this process is called "wake-up". Generally, after thousands or tens of thousands of cycles, the P_{sw} reaches its maximum, and the cell is now fully woken up. If the cycling continues, the P_{sw} may decrease over cycling, experiencing the "fatigue" process before it reaches a hard breakdown (short circuit). The wake-up/fatigue

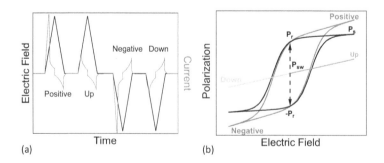

FIGURE 5.42 (a) The applied voltage waveform of PUND measurement on a ferroelectric capacitor, during which the real-time current is monitored. (b) Reconstructed P–E loop from the PUND measurement.

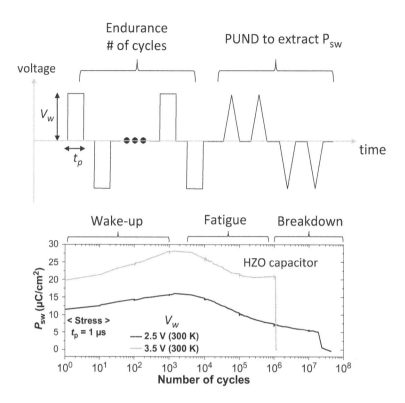

FIGURE 5.43 The endurance testing protocol and an example of the P_{sw} measured from an HZO-based ferroelectric capacitor.

process is generally attributed to existing oxygen vacancy-realignment and excessive oxygen vacancy generation. The cycling endurance of a ferroelectric capacitor depends on the write voltage amplitude/width, and less stress per pulse will improve the cycling endurance. The best-reported endurance among HZO capacitors is about 10^{10}–10^{11} cycles as of 2020.

The retention of ferroelectrics is also temperature-dependent. P_r may shrink over time at elevated temperatures, as shown in Figure 5.44(a). Similarly, the Arrhenius plot could be used to extract the activation energy, as shown in Figure 5.44(b). Essentially, P_r shrinking is reflected as the hysteresis window reduction in the P–E loop vertically.

Another unique reliability effect of ferroelectrics is the imprint effect, where the P–E loop shifts horizontally (e.g., under electrical stress or at high temperature), as shown in Figure 5.45. It is speculated that an internal bias in the ferroelectric layer is built up due to the defect dipole alignment (e.g., charged oxygen vacancies). The horizontal P-E loop shift will also reduce the sense margin as either P_r or $-P_r$ value at zero bias tends to shrink.

For NVM device integration, there are two device types of ferroelectric memories, the first one is ferroelectric random access memory (FeRAM), similar to the 1T1C DRAM's configuration; the second one is the ferroelectric field-effect transistor

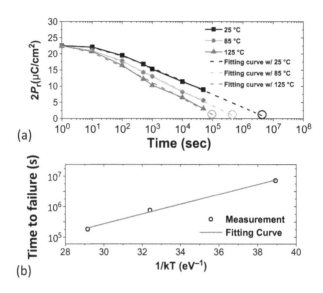

FIGURE 5.44 (a) An example of the $2P_r$ measured from an HZO-based ferroelectric capacitor at elevated temperature. (b) Arrhenius plot of the ferroelectric capacitor retention.

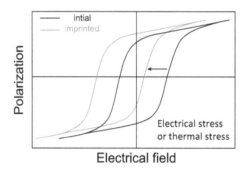

FIGURE 5.45 The imprint effect of ferroelectric material, where the P–E loop shifts horizontally under electrical stress or at high temperatures.

(FeFET), similar to the Flash transistor where the ferroelectric layer is integrated into the gate stack. These two device structures will be discussed in detail in the following sub-sections.

5.5.2   1T1C F$_e$RAM

The basic design of FeRAM involves the 1T1C bit cell, where the selection transistor's gate is controlled by WL, one of its source/drain contacts connects to BL and the other one connects to the ferroelectric capacitor. The other electrode of the ferroelectric capacitor is attached to the plate line (PL). The principle of FeRAM is similar to that of DRAM, but with one distinct feature: read-destructive. Figure 5.46 shows the representative FeRAM read operation principle and waveform. Before WL

Emerging Non-volatile Memories 175

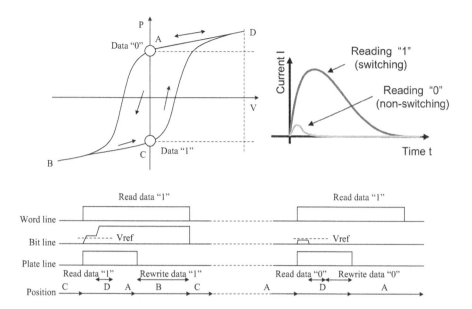

FIGURE 5.46 The representative FeRAM read operation principle and waveform. The read is destructive and a write-back is performed.

is activated, BL has an initial voltage level of zero. When WL is turned on, during the first half of the active WL pulse window, PL needs to be raised to a high voltage that is close to the write voltage. If the data is stored as "1", the cell will be in point C in the P–E hysteresis loop. The large positive PL voltage thus can flip the cell from point C to point A, resulting in large polarization switching current on the BL, which can raise the BL voltage above the reference voltage. The sense amplifier can further flip the BL voltage to V_{DD} once it is enabled. On the other hand, if the data is stored as "0", the cell will be in point A in the P–E hysteresis loop. The large positive PL voltage thus can only charge up the cell from point C to point D, resulting in a small non-switching displacement current on the BL, which cannot raise the BL voltage above the reference voltage. The sense amplifier can further flip the BL voltage to 0 once it is enabled. Since reading "1" involves a switching of the state (to point A), it is necessary to write the cell back to its original state (point C). The waveform design will make the PL ground during the second half of the active WL pulse window. Therefore, a negative voltage drop on the ferroelectric capacitor (PL = ground and BL = V_{DD}) will flip the polarization from point A back to point C (via point B). Hence, an active write-back restores the data during the read operation in FeRAM.

FeRAM with conventional perovskite ferroelectric material such as PZT has been commercially available for the niche market, for example, smart cards. A long endurance of up to 10^{14} cycles is promised by FeRAM. It should be noted that due to the destructive read, the read operation also consumes the endurance cycles in FeRAM. As >100 nm thickness is required for PZT to exhibit ferroelectricity, PZT-based FeRAM is not well scalable beyond 130 nm node. Since the discovery of the doped HfO_2 ferroelectrics (with ~10 nm thickness) in the early 2010s, the interest in

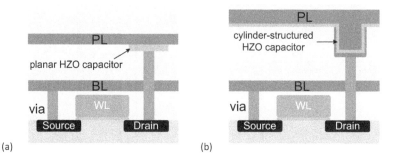

FIGURE 5.47 (a) FeRAM using a planar capacitor at 130 nm node. (b) FeRAM using a cylinder-like 3D stacked capacitor at 28 nm node.

developing a new generation of FeRAM revived. As shown in Figure 5.47(a), Sony demonstrated a 64 kb FeRAM prototype at 130 nm node with the HZO-based planar capacitor integrated at the drain's contact of the selection transistor [45], requiring a moderate 2.5 V write voltage. Fast sub-20 ns write/read speed is achieved and a relatively longer endurance (10^{11} cycles) is reported. However, to scale the HZO-based FeRAM to a more advanced node (e.g., 28 nm), a DRAM-like cylinder capacitor structure (with an aspect ratio of ~10) seems necessary to maintain the surface area for sufficient polarization charges and sense margin, as shown in Figure 5.47(b). If FeRAM can be successfully integrated at such an advanced node, it could be a competitive candidate for embedded NVM.

5.5.3 FeFET

FeFET is a three-terminal transistor structure that integrates the ferroelectric layer into the gate stack. It is unrealistic to integrate the very thick (>100 nm) perovskite ferroelectric layer into the gate stack of an advanced transistor. Practically, FeFET herein refers to the sub-10-nm-thick doped HfO_2 ferroelectric-based ones, and FeFET is typically an n-type transistor. The principle of FeFET is similar to that of Flash transistor where the threshold voltage could be modulated by the positive/negative gate voltages. The memory window is defined similarly as the distance of the high/low threshold voltages (V_T). Nevertheless, a notable difference between FeFET and Flash transistor (including both floating-gate transistor and charge-trap transistor) is that the I_D–V_G loop hysteresis direction is reversed. In FeFET, positive/negative gate voltage polarizes the domains down/up; thus, the electrons are facilitated/inhibited to create the inversion of the channel. As a result, the threshold voltage is decreased/increased and the drain current is high/low, respectively. This indicates a counter-clockwise switching in the I_D–V_G plot, as shown in Figure 5.48. On the contrary, positive/negative voltage will inject/eject electrons into/out of the gate stack in Flash transistor; thus, the threshold voltage is increased/decreased and the drain current is low/high, respectively. This indicates a clockwise switching in the I_D–V_G plot. In principle, FeFET could be also operated under MLC mode as a multi-bit memory. Since there is a distribution of the coercive fields in the multi-domain ferroelectric thin film, the gate programming pulse can vary the voltage amplitude to partially

Emerging Non-volatile Memories 177

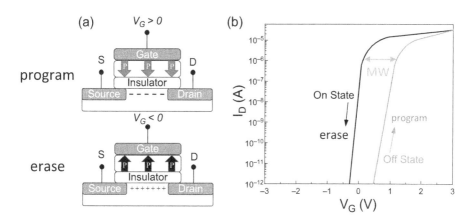

FIGURE 5.48 Schematic of the polarization in FeFET and corresponding I_D–V_G.

polarize the domains; thus, the device can be programmed into the intermediate states, as shown in Figure 5.49. It should be noted that the multilevel capability may be weakened by aggressive downscaling, especially when the gate area is reduced to dimensions that are comparable with the domain size (typically 10 nm–20 nm). Theoretically speaking, if the entire gate has only a single domain, a sharp transition in the I_D–V_G plot may result in only binary states.

The reliability effects of FeFET are generally more problematic compared to FeRAM even if they employ the same ferroelectric material. The degradation is largely because of the interfacial layer (IL) such as SiO_2 that exists between the ferroelectric layer and the silicon channel, which hampers the intrinsic reliability properties of the ferroelectric layer. The impacts of the IL can be summarized as follows. The IL capacitor and the ferroelectric capacitor forms a capacitive voltage divider; thus, most of the write voltage will undesirably drop on the IL capacitor because IL typically has a smaller dielectric constant than the ferroelectric layer. To ensure a successful write, a relatively large voltage will be applied to the gate, making the

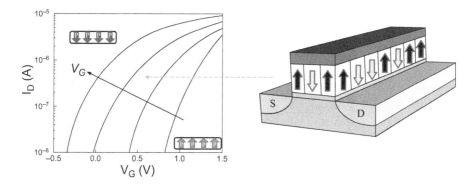

FIGURE 5.49 Partial polarization of multi-domains in FeFET and the resulting multilevel states in I_D–V_G.

write voltage to be relatively high (e.g., 3–4 V). The enhanced electric field in IL will facilitate the defect (e.g., oxygen vacancy) generation, which can serve as a trap site for capturing electrons, as shown in Figure 5.50(a). The electron trapping effect has many implications. First, it shrinks the memory window and prevents the immediate read operation after the write operation. This is because the electron trapping has an opposite effect as the polarization switching (i.e., the differences between Flash and FeFET). The electrons tend to de-trap over time, causing a relaxation process. Hence, the initially diminished memory window may open up after a while, making it impossible to read within a short period (e.g., <μs) after the write operation. Second, excessive electron trapping over cycling will make the electrons difficult to be de-trapped. If the electrons are permanently stored in the gate stack, it will make the threshold voltage always high, resulting in the erase failure in FeFET. This is the typical cause of cycling endurance degradation in FeFET, as shown in Figure 5.50(b). The low V_T state tends to merge toward the high V_T state. If without any optimization, FeFET endurance is typically in the range of 10^5–10^6 cycles. Third, the electrons accumulated at IL will induce a reverse electric field to the dipole direction within the ferroelectric layer, leading to the depolarization field that worsens the FeFET retention. To summarize, IL is known to be detrimental to FeFET performance and reliability.

There are several strategies to mitigate or eliminate the IL's impact by advanced device engineering. First, one can use IL material that has a higher dielectric constant than SiO_2, for example, SiO_xN_y with nitrogen treatment of the interface [46]. Second, one can use channel material (rather than Si) that is uneasy to form the interfacial layer, for example, oxide channel material such as W-doped In_2O_3 [47]. Third, adding a metal middle layer to separate the ferroelectric layer and IL could decouple the designs of the metal-ferroelectric-metal (MFM) ferroelectric capacitor and the underlying MOS capacitor. By tuning the area ratio between the two capacitors, the voltage drop could be mostly transferred to the ferroelectric layer [48]. With these strategies, the write voltage of FeFET can be reduced to a sub-2 V regime and the endurance of FeFET can be improved to 10^{10}–10^{12} cycles, making it attractive for many applications.

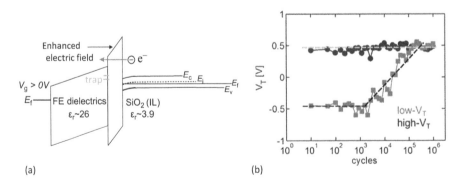

FIGURE 5.50 (a) Electron trapping at the interfacial layer due to the enhanced electric field. (b) An example of the endurance cycling of FeFET.

Another challenge for advanced FeFET integration is the variability. The V_T distribution of the programmed and erased state may have significant spread-out and possible overlap at the tail bits, as shown in Figure 5.51(a). This is a result of the domain-to-domain variation. The as-deposited ferroelectric thin film may have many microstructural grains that exist in different phases (i.e., some in ferroelectric phase and some in dielectric phase); also, the grain sizes may vary (leading to different P_r and E_c parameters within ferroelectric phase), as shown in Figure 5.51(b) [49]. Further device engineering is required to purify and unify the ferroelectric phases.

As of 2020, FeFET has been demonstrated in industrial platforms. For example, Globalfoundries adopted Si-doped HfO_2 ferroelectric gate stack and reported FeFET on high-k/metal gate platform at 28 nm node [50], and FeFET on fully depleted silicon-on-insulator (FDSOI) platform at 22 nm node [51], targeting at the embedded NVM application. Intel also reported a prototype with back-gated FeFET for embedded DRAM application (with trade-offs on longer endurance but shorter retention) [52]. The performance of these prototype devices is summarized in Table 5.7. One notable advantage of FeFET compared to other emerging NVMs is its ultra-low write energy (approaching fJ/bit) due to its electric field-driven switching mechanism. Owing to the similarity with Flash transistor, FeFET has the potential to be stacked into vertical channel 3D NAND architecture [53], offering a potential solution for ultra-high density NVM. Compared to NAND Flash, FeFET offers lower write voltage and much faster write speed. The multi-level capability of FeFET is also of interest to develop analog synaptic devices to support deep neural network acceleration [54]. To summarize, the opportunities for FeFET are wide-open.

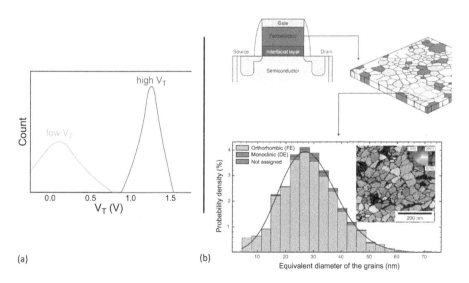

FIGURE 5.51 (a) An illustration of the threshold voltage variation in the FeFET array. (b) The grain size variation in the HZO-based ferroelectric thin film.

TABLE 5.7
The Survey of Recent FeFET Prototype Devices

Tech. Node	GF 22 nm FD-SOI	GF 28 nm HKMG	Intel back-gated (L_g = 76 nm)
	IEDM 2017	IEDM 2016	IEDM 2020
Target app.	eFLASH	eFLASH	eDRAM/DRAM
Memory window	1.5 V	1 V	0.85 V
Cell size ($F^2/\mu m^2$)	51.7 F^2/0.025 μm^2	57.4 F^2/0.045 μm^2	N/A
Write voltage (V)	4.2 V	3–4 V	1.8 V
Write pulse width (ns)	10 ns	1 μs–10 μs	10 ns
Write endurance	10^5	10^5	10^{12}
Retention	10 years @ 105°C	10 years @ 105°C	Short (>10 ms) @ 85°C

5.6 COMPUTE-IN-MEMORY

5.6.1 Principle of CIM

In recent years, the success of machine learning algorithms has motivated a wave of designing hardware accelerators for efficient implementations of deep neural network (DNN) models. Due to the abundant data volume, the primary challenge for machine learning acceleration is frequent data movement back and forth between the compute units and the memory units, namely the memory wall problem in the conventional von-Neumann architecture. To this end, compute-in-memory (CIM) is proposed as a promising paradigm since it emerges computation directly into memory sub-arrays. The operation which consumes the most part of DNN processing is the vector-matrix multiplication (VMM) between the input vector and the weight matrix, which essentially performs the multiply-and-accumulate (MAC) operation. CIM is well-positioned to accelerate VMM for machine learning workloads [55].

VMM's efficiency, if performed in the CIM manner, could be remarkably boosted by the crossbar nature of memory sub-array with perpendicular input rows and output columns, as shown in Figure 5.52. The weights of a DNN model could be mapped as the conductance of memory cells in a sub-array, while the input vector is loaded in parallel as voltage to the rows, then multiplication is done in the analog domain (i.e., the input voltage multiplied by the weight conductance), and current summation along columns is used to generate the output vector. Here, a box is used to conceptually represent the synaptic memory cell that carries the weight. The actual implementation of the synaptic memory could be a 1T1R structure (for RRAM/PCM/STT-MRAM), 1T-1F structure (for FeFET/Flash), or 8T SRAM. The weight is essentially represented by the conductance of the resistor in 1T1R, by the channel conductance of a three-terminal transistor with a tunable threshold voltage, or the channel's on/off state is determined by a latch state as in the SRAM.

Analog-to-digital converter (ADC) at the edge of the sub-array is commonly employed to convert the weighted sum (typically referred to as partial sum due to the limited sub-array size) to binary bits for subsequent digital processing (e.g., shift-and-add, accumulation, activation, and pooling). Therefore, CIM is essentially a mixed-signal scheme with analog compute core and digital peripheral processing. In

Emerging Non-volatile Memories 181

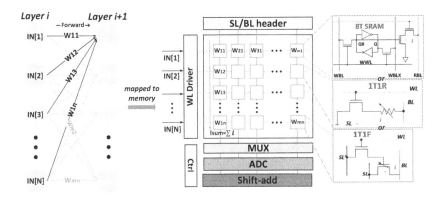

FIGURE 5.52 Schematic of compute-in-memory (CIM) paradigm. A layer of the neural network is mapped to the memory sub-arrays. Inputs are loaded in parallel as voltage to activate multiple rows, and column currents are summed up and digitalized by an analog-to-digital converter (ADC). The memory cells here could be implemented by either 1T1R, 1T1F, or 8T SRAM.

principle, VMM could be done in a fully parallel fashion if asserting all the rows and all the columns simultaneously. In practice, multiple rows/columns are partially turned on due to limited sensing resolution of ADC or mismatch between the column pitch and the ADC size.

Multi-bit weight/input/output precision could be supported in CIM. Depending on the memory cell's precision, multi-bit weight may be split into multiple cells. For example, an 8-bit weight could be represented by 2 memory cells (in 2 columns) if using 4 bits per cell. Then the outputs after ADC will need to go through a shift-and-add process to reconstruct the significance across multiple columns. The input precision could be encoded as analog voltage levels or as multiple cycles that are sequentially loaded to the rows. Due to the limited dynamic range of the input voltage that will not disturb the memory states and overhead of the digital-to-analog converter (DAC), the input precision is typically implemented by multiple cycles in practice. For example, an 8-bit input could be represented by 8 cycles with an additional shift-and-add process.

To represent both positive and negative weights, there are a couple of methods: (1) using a complementary pattern of two memory cells (e.g., in two adjacent columns) or two differential sub-arrays. Subtraction between the two columns or two sub-arrays either in an analog manner before the ADC or in a digital manner after the ADC is performed; (2) using a dummy column where the memory cells are programmed to the median level of conductance range of the multilevel cell (or using on/off states of two binary cells that are combined for average). Then the maximum positive/negative weights are mapped to the maximum/minimum conductance of the memory cell. Similarly, subtraction between the data column and the dummy column either in an analog manner before the ADC or in a digital manner after the ADC is performed; (3) using two's complement code if the memory cell is binary. The first bit is thus becoming the sign bit. CIM array performs unsigned VMM first and then utilizes the periphery to obtain correct output according to the scale and sign information.

VMM essentially performs a parallel read operation to the memory sub-array. To program the weights to initialize the DNN model for subsequent inference or to update the weights during *in-situ* training, the write operation is conducted in the memory sub-array (typically in a row-by-row fashion). A fully parallel write scheme is possible, but huge power consumption to write the entire sub-array simultaneously may be prohibitive in practice. Write-verify is commonly used to accurately tune the conductance of the memory cells for inference.

5.6.2 SYNAPTIC DEVICE PROPERTIES

The inference hardware design means that the DNN model is pre-trained by software. One-time programming is required to load the weights into memory sub-arrays. On the other hand, the training hardware design means that weights are learned on the fly during runtime. There are subtle differences between two concepts: *in-situ* training and on-chip training. It is known that the training generally requires higher weight precision than the inference only. *In-situ* training refers to the case where a single memory cell itself has sufficiently high precision to represent a weight, and the weight update occurs as the switching between intermediate states of the memory cell. The weight increment is calculated by the peripheral circuitry and is translated as the number of write pulses or the pulse amplitude. To simplify the peripheral circuit design, the identical pulsing scheme is preferred over the varying-amplitude/width pulsing scheme. On-chip training does not necessarily require a single memory cell to represent a high precision weight. Instead, it could employ multiple cells from the most significant bit (MSB) to the least significant bit (LSB). The weight increment is calculated by the peripheral circuitry and converted to binary bits to be written from MSB to LSB. In this case, on-chip training is similar to writing the data into a digital memory array.

In the following, the device properties that are of high importance for CIM are surveyed. It is noted that some requirements may be different than those for data storage. First, the device properties that matter for inference are discussed, where the read operations are intensive.

On-State Resistance: R_{on} is a critical parameter to determine the energy efficiency. As inference is read-intensive, the current sink into the ADC is directly proportional to R_{on} of individual memory cells. As a guideline to balance the power and the latency, R_{on} in the range of 100 kΩ–1 MΩ is desirable. Among all the eNVMs, FeFET offers much flexibility to meet this target range as its channel resistance could be modulated by the gate voltage bias. Moreover, emerging technologies that are specially engineered toward CIM are also encouraging. For example, a SOT-MRAM with an intentionally thicker oxide barrier could increase R_{on} to the MΩ range [56].

On/Off Ratio: Most eNVMs could achieve on/off ratios >10×, with the exception of MRAM where the on/off ratio is limited to ~2×. The small on/off ratio is not a big problem if using the dummy column or reference array to subtract the off-state current, assuming that the conductance variation is well controlled. The summed current could still be distinguished in a differential read-out manner.

Multilevel: Multi-bit is desired for higher integration density. SRAM/MRAM is binary in nature, while other eNVMs generally offer multi-bit per cell, and 2–3 bits

Emerging Non-volatile Memories

per cell is commonly achievable. The multi-bit cell is generally realized with partial switching of the materials. As shown in Figure 5.53, RRAM relies on the modulation of the filament strength, PCM relies on the modulation of the amorphous volume, and FeFET relies on the modulation of the partially flipped domains. With special engineering, 5 bits per cell FeFET [54], 5 bits per cell RRAM [57], and 10 bits per cell PCM [58] have been demonstrated. It is known that inference requires less precision than training from the algorithm's point of view. Typically, 2–3 bits per cell is sufficient for inference while 7–8 bits per cell is required for *in-situ* training.

Retention: For inference, the stability of weights over time is of high importance, which is translated to the retention of eNVMs conductance. Mechanisms that affect the retention include the short-term relaxation (after initial programming), the long-term drift (especially at elevated temperature), and the read disturb (with stress voltage). The drift of conductance may have different scenarios as shown in Figure 5.54(a). DNN+NeuroSim [59] is a device-to-system benchmark framework, which could be leveraged to estimate the inference accuracy degradation. Figure 5.54(b) shows the inference accuracy of CIFAR-10 image recognition as a function of time, while the conductance is assumed to drift toward different final states (from −1 to 1, according to the algorithm weight range), or random drift, with a rate of conductance drift by

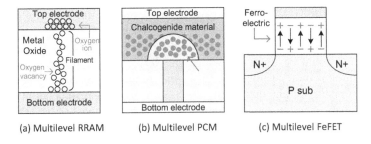

FIGURE 5.53 Landscape of multi-bit synaptic memories.

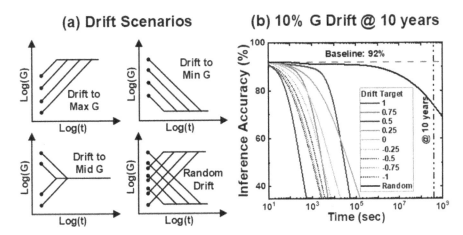

FIGURE 5.54 (a) Various scenarios of conductance drift with different targets. (b) CIFAR-10 inference accuracy vs. time considering conductance drift.

10% over 10 years. The results show that, in scenarios with fixed drift directions, drifting to maximum or minimal states degrades the accuracy faster than drifting to the middle states, while the random drift is the most robust for maintaining inference accuracy. More experimental characterizations are still needed to fully capture the stability of intermediate states of the eNVMs.

Next, the device properties that matter for training are discussed, where the write operations should be taken into consideration.

Write Voltage: Write voltage of many eNVMs (including RRAM/PCM/FeFET) are still in the range of 2–3 V, which is higher than the typical logic process supply voltage (0.7 V–1 V). Therefore, the level shifter that converts the voltage domains is needed. Since the level shifter employs I/O transistors, it may take a significant portion of the area. STT-MRAM/SOT-MRAM naturally offers low write voltage (~1 V or below).

Endurance: Training requires frequent write operations to the memory cells; however, it is often thought that endurance requirement is prohibitive for today's eNVMs for training. It turns out that endurance of ~10^6 cycles is sufficient for most of the image recognition training workloads if using batch-mode training. It is also worthwhile to note that the weight update is incremental, and the switching of eNVMs is thus not fully between the highest conductance and the lowest conductance. The intermediate switching could relax the endurance requirement. For example, a typical RRAM cell could switch ~10^5 cycles in the full switching while it could switch ~10^{11} cycles in the intermediate switching [60].

Variation: Device-to-device variation is tolerable to some extent in the iterative training, and could be tightened with write-verify for the inference. Cycle-to-cycle variation is more problematic for *in-situ* training if the randomness exceeds the deterministic weight update direction. However, a small amount of noise injected into the training actually could improve the robustness of the DNN model against variations in the subsequent inference, as the system could be converged to a local minima where its energy landscape of loss function is shallow (thus insensitive to variations).

Asymmetry/Nonlinearity: It is well known that asymmetry and nonlinearity in the weight update curve (conductance vs. # identical programming pulses) are the most critical properties that may significantly degrade *in-situ* training accuracy. Figure 5.55(a)–(c) shows the ideal weight update curves and the behavioral model that labels nonlinearity factor from 0 to 8 and asymmetry as +/−. Figure 5.55(d) shows that the *in-situ* training accuracy quickly degrades with the nonlinearity factor if the device exhibits asymmetry [61]. Algorithmic tricks (Tiki-Taka in weight update [62], and momentum in weight update [63]) suggest some relaxed requirements on linearity and symmetry. However, realizing software-comparable *in-situ* training accuracy is still quite challenging for a complex dataset. Alternative hybrid-precision synapse designs [64, 65] were recently proposed, which leveraged a volatile capacitor for symmetric and linear fine-tuning and a non-volatile memory component for coarse tuning, thereby approaching the software baseline training accuracy. It should be pointed out that nonlinearity/asymmetry is not a problem for inference, as write-verify could be used to enforce a linear mapping between weights in the algorithm and conductance of the eNVM devices.

Emerging Non-volatile Memories

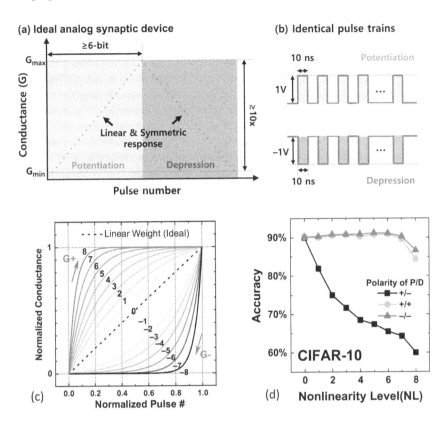

FIGURE 5.55 (a)–(b) The ideal weight update curve, (c) the behavioral model that labels nonlinearity factor from 0 to 8 and asymmetry as +/–, and (d) the resulting inference accuracy degradation for the CIFAR-10 dataset.

5.6.3 CIM Prototype Chips

Among all the memory technologies, SRAM is a mature candidate for CIM, with state-of-the-art process availability at 5 nm and beyond. However, leakage power will be the main concern for large-capacity SRAM arrays. 6T SRAM is the most compact bit cell, but the read disturb is present when multiple rows are activated. Lowering WL voltage will help, but the number of rows that can be turned on is rather limited. 8T SRAM decouples read-write paths and thus offers more flexibility for CIM design. Foundry typically offers 6T and 8T SRAM compact design rules. If further changing the bit cell (e.g., adding more transistors or changing the interconnect routing), the logic design rule that is handcrafted must be employed, resulting in much lower density (2×–4×) from the layout's perspective. In the legacy technology node (e.g., 65 nm), there is more room to modify the bit cell and reroute interconnect, but in the advanced technology node (e.g., 28 nm or beyond), foundry typically does not offer exceptions for making these changes.

As of 2020, macro-level demonstrations of CIM prototype chips have been shown in SRAM and RRAM platforms for inference. 7 nm 8T-SRAM [66] and 22 nm

RRAM (with 1 bit per cell) [67] have been integrated with peripheral ADC circuit to enable parallel compute. Because most of the accelerator designs support fixed-precision computation (but with different precisions for different designs), it is suggested to normalize the performance metrics to the same input/weight precision for a fair comparison. Here, 8-bit by 8-bit MAC is defined as two operations.

Figure 5.56 shows the survey of the reported performance (in terms of terra-operations-per-second, TOPS) versus the power in (terms of W) among the CIM prototypes and other accelerator platforms including the GPU, digital application-specific-integrated-circuits (ASIC) (such as tensor-processing-unit, TPU [68], and its variants). In this chart, the star points are projected by the DNN+NeuroSim simulation. The straight lines are the contours of iso-energy efficiency (in terms of TOPS/W). A few observations could be made. First, SRAM CIM at the leading-edge node (e.g., 7 nm) achieves the best energy efficiency (approaching 100 TOPS/W for 8-bit by 8-bit MAC), and it is 100× to 10× higher than that of GPU or TPU variants. RRAM CIM at a cost-effective node (e.g., 22 nm) could also approach 10 TOPS/W. The reason that RRAM CIM has lower energy efficiency than SRAM is partly due to the unoptimized R_{on} (a few kΩ as opposed to the desired >100 kΩ). With this regard, FeFET CIM could achieve comparable or even better energy efficiency than the SRAM counterpart, but the FeFET CIM prototype is yet to be demonstrated as of 2020. Second, the CIM prototype data in the plot is usually collected from a small macro with limited memory capacity; thus, the throughput is typically small with low power consumption. If a scalable architecture is used to duplicate multi-macros for a larger-scale system, the throughput and power of the CIM prototypes could be proportional to the number of macros to the first order. Hence, CIM could still stay with the same straight line, maintaining its superior energy efficiency. It should be pointed

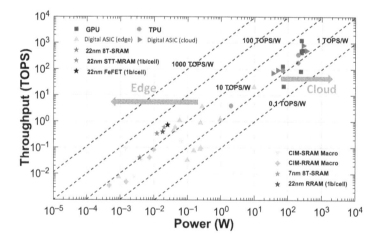

FIGURE 5.56 The survey of the reported performance (in terms of TOPS) versus the power in (terms of W) among the CIM prototypes and other accelerator platforms including the GPU, ASIC such as TPU and its variants. In this chart, the star points are projected by the DNN+NeuroSim simulation.

Emerging Non-volatile Memories

out though, CIM generally targets at edge applications with a constrained power budget, while GPU/TPU aims at high-performance cloud computing.

Figure 5.57 shows the relationship between the compute efficiency in terms of TOPS/mm^2 versus the energy efficiency in terms of TOPS/W. In this chart, the straight lines represent the power density in terms of W/mm^2. Most CIM designs, if using mature technology nodes (e.g., 22 nm or above), will have a low power density. The 7 nm SRAM CIM prototype has a high compute efficiency because of the small cell area in the absolute area, but at the same time, it results in high power density.

Despite the promises of better energy efficiency, the CIM paradigm has challenges in the following aspects [69]. First, the mixed-signal compute scheme is inherently prone to noises/variations. The inference accuracy is typically lower than the digital compute counterpart even with the same input/weight precision. The spatial variations originate from transistor mismatch, and eNVM conductance variation, etc. The temporal variations come from thermal noises, and instability/aging of eNVM conductance, etc. Second, the relatively low R_{on} of eNVM devices may result in a large column current when being parallelly accessed. Thus, a sized-up MUX is needed to avoid significant voltage drop across the pass-gate transistor. The sized-up MUX takes a significant portion of the area and contributes to large load capacitance. In addition, the large column current will sink to the ADC, causing excessive power consumption at ADC. Third, the relatively high write voltage of some eNVM devices such as RRAM, PCM, and FeFET will require using large I/O transistors for the level shifter, causing significant area overhead. Improvements in throughput and energy efficiency are highly desirable for real-time decision-making and power-constrained edge devices. Potential drawbacks of the CIM paradigm, such as degraded accuracy, may be tolerable for certain types of edge applications and could be mitigated by algorithmic techniques such as variation-aware training [70]. The capability of

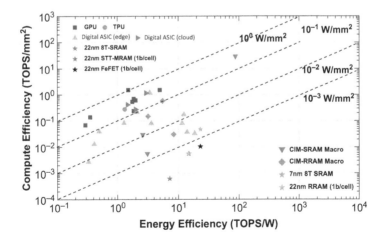

FIGURE 5.57 The survey of the reported compute efficiency (in terms of TOPS/mm^2) versus the energy efficiency in (terms of TOPS/W) among the CIM prototypes and other accelerator platforms including the GPU, ASIC such as TPU and its variants. In this chart, the star points are projected by the DNN+NeuroSim simulation.

processing information locally at the sensor/camera frontend is beneficial for saving bandwidth and energy to send data wirelessly back to the cloud. Security/privacy is another motivation to develop smart edge platforms as users may be reluctant to share personal data. Therefore, CIM is a compelling paradigm that demands further research and investment.

NOTES

1. In some literature, it is also referred to as memristor.
2. SOT-MRAM is a three-terminal device.
3. For memory operation, capacitance is under the context of large-signal swing. It should not be referred to as the small-signal capacitance under AC stimuli.
4. $\ln(10) = 2.3$, is the coefficient factor to covert ln to $\log_{10}$.
5. Subscript x for HfO_x and TaO_x suggest these oxides are in sub-stoichiometry.
6. 1 ppm is the probability of one part per one million parts.

REFERENCES

[1] S. Yu, P.-Y. Chen, "Emerging memory technologies: recent trends and prospects," *IEEE Solid-State Circuits Magazine*, vol. 8, no. 2, pp. 43–56, 2016. doi: 10.1109/MSSC.2016.2546199

[2] X. Peng, R. Madler, P.-Y. Chen, S. Yu, "Cross-point memory design challenges and survey of selector device characteristics," *Journal of Computational Electronics*, vol. 16, no. 4, pp. 1167–1174, 2017. doi: 10.1007/s10825-017-1062-z

[3] J.-J. Huang, Y.-M. Tseng, W.-C. Luo, C.-W. Hsu, T.-H. Hou, "One selector-one resistor (1S1R) crossbar array for high-density flexible memory applications," *IEEE International Electron Devices Meeting (IEDM)*, 2011, pp. 31.7.1–31.7.4. doi: 10.1109/IEDM.2011.6131653

[4] W. Lee, J. Park, J. Shin, J. Woo, S. Kim, G. Choi, S. Jung, S. Park, D. Lee, E. Cha, H.D. Lee, S.G. Kim, S. Chung, H. Hwang, "Varistor-type bidirectional switch (J_{MAX}>10^{7}A/cm^2, selectivity~10^4) for 3D bipolar resistive memory arrays," *IEEE Symposium on VLSI Technology*, 2012, pp. 37–38. doi: 10.1109/VLSIT.2012.6242449

[5] L. Zhang, B. Govoreanu, A. Redolfi, D. Crotti, H. Hody, V. Paraschiv, S. Cosemans, C. Adelmann, T. Witters, S. Clima, Y.-Y. Chen P. Hendrickx, D.J. Wouters, G. Groeseneken, M. Jurczak, "High-drive current (>1MA/cm^2) and highly nonlinear (>10^3) TiN/amorphous-Silicon/TiN scalable bidirectional selector with excellent reliability and its variability impact on the 1S1R array performance," *IEEE International Electron Devices Meeting (IEDM)*, 2014, pp. 6.8.1–6.8.4. doi: 10.1109/IEDM.2014.7047000

[6] Q. Luo, X. Xu, H. Lv, T. Gong, S. Long, Q. Liu, H. Sun, L. Li, N. Lu, M. Liu, "Fully BEOL compatible TaOx-based selector with high uniformity and robust performance," *IEEE International Electron Devices Meeting (IEDM)*, 2016, pp. 11.7.1–11.7.4. doi: 10.1109/IEDM.2016.7838399

[7] K. Virwani, G.W. Burr, R.S. Shenoy, C.T. Rettner, A. Padilla, T. Topuria, P.M. Rice, G. Ho, R.S. King, K. Nguyen, A.N. Bowers, M. Jurich, et al., "Sub-30nm scaling and high-speed operation of fully-confined access-devices for 3D crosspoint memory based on mixed-ionic-electronic-conduction (MIEC) materials," *IEEE International Electron Devices Meeting (IEDM)*, 2012, pp. 2.7.1–2.7.4. doi: 10.1109/IEDM.2012.6478967

[8] S.G. Kim, T.J. Ha, S. Kim, J.Y. Lee, K.W. Kim, J.H. Shin, Y.T. Park, S.P. Song, et al., "Improvement of characteristics of NbO_2 selector and full integration of 4F^2 2x-nm tech 1S1R ReRAM," *IEEE International Electron Devices Meeting (IEDM)*, 2015, pp. 10.3.1–10.3.4. doi: 10.1109/IEDM.2015.7409668

[9] D. Kau, S. Tang, I.V. Karpov, R. Dodge, B. Klehn, J. Kalb, J. Strand, A. Diaz, N. Leung, J. Wu, S. Lee, et al., "A stackable cross point phase change memory," *IEEE International Electron Devices Meeting (IEDM)*, 2009, pp. 1–4. doi: 10.1109/IEDM.2009.5424263

[10] M.-J. Lee, D. Lee, H. Kim, H.-S. Choi, J.-B. Park, H.G. Kim, Y.-K. Cha, U.-I. Chung, I.-K. Yoo, K. Kim, "Highly-scalable threshold switching select device based on chaclogenide glasses for 3D nanoscaled memory arrays," *IEEE International Electron Devices Meeting (IEDM)*, 2012, pp. 2.6.1–2.6.3. doi: 10.1109/IEDM.2012.6478966

[11] Y. Koo, K. Baek, H. Hwang, "Te-based amorphous binary OTS device with excellent selector characteristics for X-point memory applications," *IEEE Symposium on VLSI Technology*, 2016, pp. 1–2. doi: 10.1109/VLSIT.2016.7573389

[12] S.H. Jo, T. Kumar, S. Narayanan, W.D. Lu, H. Nazarian, "3D-stackable crossbar resistive memory based on field assisted superlinear threshold (FAST) selector," *IEEE International Electron Devices Meeting (IEDM)*, 2014, pp. 6.7.1–6.7.4. doi: 10.1109/IEDM.2014.7046999

[13] Q. Luo, X. Xu, H. Liu, H. Lv, T. Gong, S. Long, Q. Liu, H. Sun, W. Banerjee, L. Li, N. Lu, M. Liu, "Cu BEOL compatible selector with high selectivity (>10^7), extremely low off-current (~pA) and high endurance (>10^{10})," *IEEE International Electron Devices Meeting (IEDM)*, 2015, pp. 10.4.1–10.4.4. doi: 10.1109/IEDM.2015.7409669

[14] F. Arnaud, P. Ferreira, F. Piazza, A. Gandolfo, P. Zuliani, P. Mattavelli, E. Gomiero, et al., "High density embedded PCM cell in 28nm FDSOI technology for automotive micro-controller applications," *IEEE International Electron Devices Meeting (IEDM)*, 2020, pp. 24.2.1–24.2.4. doi: 10.1109/IEDM13553.2020.9371934.

[15] F. Arnaud, P. Zuliani, J.P. Reynard, A. Gandolfo, F. Disegni, P. Mattavelli, E. Gomiero, et al., "Truly innovative 28nm FDSOI technology for automotive micro-controller applications embedding 16MB phase change memory," *IEEE International Electron Devices Meeting (IEDM)*, 2018, pp. 18.4.1–18.4.4. doi: 10.1109/IEDM.2018.8614595

[16] J.Y. Wu, Y.S. Chen, W.S. Khwa, S.M. Yu, T.Y. Wang, J.C. Tseng, Y.D. Chih, C.H. Diaz, "A 40nm low-power logic compatible phase change memory technology," *IEEE International Electron Devices Meeting (IEDM)*, 2018, pp. 27.6.1–27.6.4. doi: 10.1109/IEDM.2018.8614513

[17] Y. Choi, I. Song, M.-H. Park, H. Chung, S. Chang, B. Cho, J. Kim, et al., "A 20nm 1.8 V 8Gb PRAM with 40MB/s program bandwidth," *IEEE International Solid-State Circuits Conference (ISSCC)*, 2012, pp. 46–48. doi: 10.1109/ISSCC.2012.6176872

[18] M. Zhu, K. Ren, Z. Song, "Ovonic threshold switching selectors for three-dimensional stackable phase-change memory," *MRS Bulletin*, vol. 44, no. 9, pp. 715–720, 2019. doi: 10.1557/mrs.2019.206

[19] ReverseEngineeringReporton3DX-pointbyTechInsights,https://www.techinsights.com/blog/intel-3d-xpoint-memory-die-removed-intel-optanetm-pcm-phase-change-memory

[20] B. Govoreanu, G.S. Kar, Y. Chen, V. Paraschiv, S. Kubicek, A. Fantini, I.P. Radu, L. Goux, S. Clima, R. Degraeve, N. Jossart, O. Richard, T. Vandeweyer, K. Seo, P. Hendrickx, G. Pourtois, H. Bender, L. Altimime, D.J. Wouters, J.A. Kittl, M. Jurczak, "10×10nm^2 Hf/HfO$_x$ crossbar resistive RAM with excellent performance, reliability and low-energy operation," *IEEE International Electron Devices Meeting (IEDM)*, 2011, pp. 31.6.1–31.6.4. doi: 10.1109/IEDM.2011.6131652

[21] Y.Y. Chen, B. Govoreanu, L. Goux, R. Degraeve, Andrea Fantini, G.S. Kar, D.J. Wouters, G. Groeseneken, J.A. Kittl, M. Jurczak, L. Altimime, "Balancing SET/RESET pulse for > 1E10 endurance in HfO$_2$/Hf 1T1R bipolar RRAM," *IEEE Transactions on Electron Devices*, vol. 59, no. 12, pp. 3243–3249, 2012. doi: 10.1109/TED.2012.2218607

[22] Y.Y. Chen, R. Degraeve, S. Clima, B. Govoreanu, L. Goux, A. Fantini, G.S. Kar, G. Pourtois, G. Groeseneken, D.J. Wouters, M. Jurczak, "Understanding of the endurance

failure in scaled HfO$_2$-based 1T1R RRAM through vacancy mobility degradation," *IEEE International Electron Devices Meeting (IEDM)*, 2012, pp. 20.3.1–20.3.4. doi: 10.1109/IEDM.2012.6479079

[23] S.-S. Sheu, M.-F. Chang, K.-F. Lin, C.-W. Wu, Y.-S. Chen, P.-F. Chiu, C.-C. Kuo, Y.-S. Yang, P.-C. Chiang, W.-P. Lin, C.-H. Lin, H.-Y. Lee, P.-Y. Gu, S.-M. Wang, F.T. Chen, K.-L. Su, C.-H. Lien, K.-H. Cheng, H.-T. Wu, T.-K. Ku, M.-J. Kao, M.-J. Tsai, "A 4Mb embedded SLC resistive-RAM macro with 7.2ns read-write random-access time and 160ns MLC-access capability," *IEEE International Solid-State Circuits Conference (ISSCC)*, 2011, pp. 200–202. doi: 10.1109/ISSCC.2011.5746281

[24] A. Kawahara, R. Azuma, Y. Ikeda, K. Kawai, Y. Katoh, K. Tanabe, T. Nakamura, Y. Sumimoto, N. Yamada, N. Nakai, S. Sakamoto, Y. Hayakawa, K. Tsuji, S. Yoneda, A. Himeno, K. Origasa, K. Shimakawa, T. Takagi, T. Mikawa, K. Aono, "An 8Mb multi-layered cross-point ReRAM macro with 443MB/s write throughput," *IEEE International Solid-State Circuits Conference (ISSCC)*, 2012, pp. 432–434. doi: 10.1109/ISSCC.2012.6177078

[25] C. Ho, S.-C. Chang, C.-Y. Huang, Y.-C. Chuang, S.-F. Lim, M.-H. Hsieh, S.-C. Chang, H.-H. Liao, "Integrated HfO$_2$-RRAM to achieve highly reliable, greener, faster, cost-effective, and scaled devices," *IEEE International Electron Devices Meeting (IEDM)*, 2017, pp. 2.6.1–2.6.4. doi: 10.1109/IEDM.2017.8268314

[26] C.-C. Chou, Z.-J. Lin, P.-L. Tseng, C.-F. Li, C.-Y. Chang, W.-C. Chen, Y.-D. Chih, T.-Y.J. Chang, "An N40 256K×44 embedded RRAM macro with SL-precharge SA and low-voltage current limiter to improve read and write performance," *IEEE International Solid-State Circuits Conference (ISSCC)*, 2018, pp. 478–480. doi: 10.1109/ISSCC.2018.8310392

[27] C.F. Yang, C.-Y. Wu, M.-H. Yang, W. Wang, M.-T. Yang, T.-C. Chien, V. Fan, et al., "Industrially applicable read disturb model and performance on Mega-bit 28nm embedded RRAM," *IEEE Symposium on VLSI Technology*, 2020, pp. 1–2. doi: 10.1109/VLSITechnology18217.2020.9265060

[28] C.-C. Chou, Z.-J. Lin, C.-A. Lai, C.-I. Su, P.-L. Tseng, W.-C. Chen, W.-C. Tsai, et al., "A 22nm 96K×144 RRAM macro with a self-tracking reference and a low ripple charge pump to achieve a configurable read window and a wide operating voltage range," *IEEE Symposium on VLSI Circuits*, 2020, pp. 1–2. doi: 10.1109/VLSICircuits18222.2020.9163014

[29] P. Jain, U. Arslan, M. Sekhar, B.C. Lin, L. Wei, T. Sahu, J. Alzate-Vinasco, et al., "A 3.6 Mb 10.1 Mb/mm^2 embedded non-volatile ReRAM macro in 22nm FinFET technology with adaptive forming/set/reset schemes yielding down to 0.5 V with sensing time of 5ns at 0.7 V," *IEEE International Solid-State Circuits Conference (ISSCC)*, 2019, pp.212–214. doi: 10.1109/ISSCC.2019.8662393

[30] T.-Y. Liu, T.H. Yan, R. Scheuerlein, Y. Chen, J.K. Lee, G. Balakrishnan, G. Yee, H. Zhang, A. Yap, J. Ouyang, et al., "A 130.7mm^2 2-layer 32Gb ReRAM memory device in 24nm technology," *IEEE International Solid-State Circuits Conference (ISSCC)*, 2013, pp. 210–211. doi: 10.1109/ISSCC.2013.6487703

[31] R. Fackenthal, M. Kitagawa, W. Otsuka, K. Prall, D. Mills, K. Tsutsui, J. Javanifard, et al., "A 16Gb ReRAM with 200MB/s write and 1GB/s read in 27nm technology," *IEEE International Solid-State Circuits Conference (ISSCC)*, 2014, pp. 338–339. doi: 10.1109/ISSCC.2014.6757460

[32] W.J. Gallagher, S.S.P. Parkin, "Development of the magnetic tunnel junction MRAM at IBM: from first junctions to a 16-Mb MRAM demonstrator chip," *IBM Journal of Research and Development*, vol. 50, no. 1, pp. 5–23, 2006. doi: 10.1147/rd.501.0005

[33] V.B. Naik, K. Yamane, T.Y. Lee, J. Kwon, R. Chao, J.H. Lim, N.L. Chung, et al., "JEDEC-qualified highly reliable 22nm FD-SOI embedded MRAM for low-power

industrial-grade, and extended performance towards automotive-grade-1 applications," *IEEE International Electron Devices Meeting (IEDM)*, 2020, pp. 11.3.1–11.3.4. doi: 10.1109/IEDM13553.2020.9371935

[34] O. Golonzka, J.-G. Alzate, U. Arslan, M. Bohr, P. Bai, J. Brockman, B. Buford, et al., "MRAM as embedded non-volatile memory solution for 22FFL FinFET technology," *IEEE International Electron Devices Meeting (IEDM)*, 2018, pp. 18.1.1–18.1.4. doi: 10.1109/IEDM.2018.8614620

[35] Y.-C. Shih, C.-F. Lee, Y.-A. Chang, P.-H. Lee, H.-J. Lin, Y.-L. Chen, C.-P. Lo, et al., "A reflow-capable, embedded 8Mb STT-MRAM macro with 9ns read access time in 16nm FinFET logic CMOS process," *IEEE International Electron Devices Meeting (IEDM)*, 2020, pp. 11.4.1–11.4.4. doi: 10.1109/IEDM13553.2020.9372115

[36] Y.J. Song, J.H. Lee, S.H. Han, H.C. Shin, K.H. Lee, K. Suh, D.E. Jeong, et al., "Demonstration of highly manufacturable STT-MRAM embedded in 28nm logic," *IEEE International Electron Devices Meeting (IEDM)*, 2018, pp. 18.2.1–18.2.4. doi: 10.1109/IEDM.2018.8614635

[37] T.Y. Lee, K. Yamane, Y. Otani, D. Zeng, J. Kwon, J.H. Lim, V.B. Naik, et al., "Advanced MTJ stack engineering of STT-MRAM to realize high speed applications," *IEEE International Electron Devices Meeting (IEDM)*, 2020, pp. 11.6.1–11.6.4. doi: 10.1109/IEDM13553.2020.9372015

[38] D. Edelstein, M. Rizzolo, D. Sil, A. Dutta, J. DeBrosse, M. Wordeman, A. Arceo, et al., "A 14 nm embedded STT-MRAM CMOS technology," *IEEE International Electron Devices Meeting (IEDM)*, 2020, pp. 11.5.1–11.5.4. doi: 10.1109/IEDM13553.2020.9371922

[39] J.G. Alzate, U. Arslan, P. Bai, J. Brockman, Y.-J. Chen, N. Das, K. Fischer, et al., "2 MB array-level demonstration of STT-MRAM process and performance towards L4 cache applications," *IEEE International Electron Devices Meeting (IEDM)*, 2019, pp. 2.4.1–2.4.4. doi: 10.1109/IEDM19573.2019.8993474

[40] K. Garello, F. Yasin, H. Hody, S. Couet, L. Souriau, S.H. Sharifi, J. Swerts, R. Carpenter, et al., "Manufacturable 300mm platform solution for field-free switching SOT-MRAM," *IEEE Symposium on VLSI Circuits*, 2019, pp. T194–T195. doi: 10.23919/VLSIC.2019.8778100

[41] M. Gupta, M. Perumkunnil, K. Garello, S. Rao, F. Yasin, G.S. Kar, A. Furnemont, "High-density SOT-MRAM technology and design specifications for the embedded domain at 5nm node," *IEEE International Electron Devices Meeting (IEDM)*, 2020, pp. 24.5.1–24.5.4. doi: 10.1109/IEDM13553.2020.9372068

[42] H. Honjo, T.V.A. Nguyen, T. Watanabe, T. Nasuno, C. Zhang, T. Tanigawa, S. Miura, et al., "First demonstration of field-free SOT-MRAM with 0.35 ns write speed and 70 thermal stability under 400 °C thermal tolerance by canted SOT structure and its advanced patterning/SOT channel technology," *IEEE International Electron Devices Meeting (IEDM)*, 2019, pp. 28.5.1–28.5.4. doi: 10.1109/IEDM19573.2019.8993443

[43] A.I. Khan, K. Chatterjee, B. Wang, S. Drapcho, L. You, C. Serrao, S.R. Bakaul, R. Ramesh, S. Salahuddin, "Negative capacitance in a ferroelectric capacitor," *Nature Materials*, vol. 14, no. 2, p. 182, 2015. doi: 10.1038/nmat4148

[44] Z. Wang, J. Hur, N. Tasneem, W. Chern, S. Yu, A.I. Khan, "Extraction of Preisach model parameters for fluorite-structure ferroelectrics and antiferroelectrics," *Scientific Reports*, vol. 11, p. 12474, 2021. doi: 10.1038/s41598-021-91492-w

[45] J. Okuno, T. Kunihiro, K. Konishi, H. Maemura, Y. Shute, F. Sugaya, M. Materano, et al., "SoC compatible 1T1C FeRAM memory array based on ferroelectric $Hf_{0.5}Zr_{0.5}O_2$," *IEEE Symposium on VLSI Technology*, 2020, pp. 1–2, doi: 10.1109/VLSITechnology18217.2020.9265063

[46] A.J. Tan, Y.H. Liao, L.C. Wang, N. Shanker, J.H. Bae, C. Hu, S. Salahuddin, "Ferroelectric HfO$_2$ memory transistors with high-k interfacial layer and write endurance exceeding 10^{10} cycles," *IEEE Electron Device Letters*, vol. 42, no. 7, pp. 994–997, 2021. doi: 10.1109/LED.2021.3083219.

[47] S. Dutta, H. Ye, W. Chakraborty, Y.-C. Luo, M. San Jose, B. Grisafe, A. Khanna, I. Lightcap, S. Shinde, S. Yu, S. Datta, "Monolithic 3D integration of high endurance multi-bit ferroelectric FET for accelerating compute-in-memory," *IEEE International Electron Devices Meeting (IEDM)*, 2020, pp. 36.4.1–36.4.4. doi: 10.1109/IEDM13553.2020.9371974

[48] K. Ni, J.A. Smith, B. Grisafe, T. Rakshit, B. Obradovic, J.A. Kittl, M. Rodder, S. Datta, "SoC logic compatible multi-bit FeMFET weight cell for neuromorphic applications," *IEEE International Electron Devices Meeting (IEDM)*, 2018, pp. 13.2.1–13.2.4. doi: 10.1109/IEDM.2018.8614496

[49] M. Lederer, T. Kampfe, R. Olivo, D. Lehninger, C. Mart, S. Kirbach, T. Ali, P. Polakowski, L. Roy, K. Seidel, "Local crystallographic phase detection and texture mapping in ferroelectric Zr doped HfO$_2$ films by transmission-EBSD," *Applied Physics Letters*, vol. 115, p. 222902, 2019. doi: 10.1063/1.5129318

[50] M. Trentzsch, S. Flachowsky, R. Richter, J. Paul, B. Reimer, D. Utess, S. Jansen, H. Mulaosmanovic, S. Muller, et al., "A 28nm HKMG super low power embedded NVM technology based on ferroelectric FETs," *IEEE International Electron Devices Meeting (IEDM)*, 2016, pp. 11.5.1–11.5.4. doi: 10.1109/IEDM.2016.7838397

[51] S. Dunkel, M. Trentzsch, R. Richter, P. Moll, C. Fuchs, O. Gehring, M. Majer, S. Wittek, B. Muller, et al., "A FeFET based super-low-power ultra-fast embedded NVM technology for 22nm FDSOI and beyond," *IEEE International Electron Devices Meeting (IEDM)*, 2017, pp. 19.7.1–19.7.4. doi: 10.1109/IEDM.2017.8268425

[52] A. Sharma, B. Doyle, H.J. Yoo, I.-C. Tung, J. Kavalieros, M.V. Metz, M. Reshotko, et al., "High speed memory operation in channel-last, back-gated ferroelectric transistors," *IEEE International Electron Devices Meeting (IEDM)*, 2020, pp. 18.5.1–18.5.4, doi: 10.1109/IEDM13553.2020.9371940.

[53] K. Florent, S. Lavizzari, L. Di Piazza, M. Popovici, E. Vecchio, G. Potoms, G. Groeseneken, J. Van Houdt, "First demonstration of vertically stacked ferroelectric Al doped HfO$_2$ devices for NAND applications," *IEEE Symposium on VLSI Technology*, 2017, pp. T158–T159. doi: 10.23919/VLSIT.2017.7998162

[54] M. Jerry, P.-Y. Chen, J. Zhang, P. Sharma, K. Ni, S. Yu, S. Datta, "Ferroelectric FET analog synapse for acceleration of deep neural network training," *IEEE International Electron Devices Meeting (IEDM)*, 2017, pp. 6.2.1–6.2.4. doi: 10.1109/IEDM.2017.8268338

[55] S. Yu, H. Jiang, S. Huang, X. Peng, A. Lu, "Compute-in-memory chips for deep learning: recent trends and prospects," *IEEE Circuits and Systems Magazine*, vol. 21, no. 3, pp. 31–56, 2021. doi: 10.1109/MCAS.2021.3092533

[56] J. Doevenspeck, K. Garello, B. Verhoef, R. Degraeve, S. Van Beek, D. Crotti, F. Yasin, S. Couet, G. Jayakumar, I.A. Papistas, P. Debacker, R. Lauwereins, W. Dehaene, G.S. Kar, S. Cosemans, A. Mallik, D. Verkest, "SOT-MRAM based analog in-memory computing for DNN inference," *IEEE Symposium on VLSI Technology*, 2020, pp. 1–2. doi: 10.1109/VLSITechnology18217.2020.9265099

[57] P. Yao, H. Wu, B. Gao, J. Tang, Q. Zhang, W. Zhang, J.J. Yang, H. Qian, "Fully hardware-implemented memristor convolutional neural network," *Nature*, vol. 577, no. 7792, pp. 641–646, 2020. doi: 10.1038/s41586-020-1942-4

[58] W. Kim, R.L. Bruce, T. Masuda, G.W. Fraczak, N. Gong, P. Adusumilli, S. Ambrogio, H. Tsai, J. Bruley, J.-P. Han, M. Longstreet, "Confined PCM-based analog synaptic

devices offering low resistance-drift and 1000 programmable states for deep learning," *IEEE Symposium on VLSI Technology*, 2019, pp. T66–T67. doi: 10.23919/VLSIT.2019.8776551

[59] X. Peng, S. Huang, Y. Luo, X. Sun, S. Yu, "DNN+NeuroSim: an end-to-end benchmarking framework for compute-in-memory accelerators with versatile device technologies," *IEEE International Electron Devices Meeting (IEDM)*, 2019, pp. 32.5.1–32.5.4. doi: 10.1109/IEDM19573.2019.8993491 Online available at https://github.com/neurosim

[60] M. Zhao, H. Wu, B. Gao, X. Sun, Y. Liu, P. Yao, Y. Xi, X. Li, Q. Zhang, K. Wang, S. Yu, H. Qian, "Characterizing endurance degradation of incremental switching in analog RRAM for neuromorphic systems," *IEEE International Electron Devices Meeting (IEDM)*, 2018, pp. 20.2.1–20.2.4. doi: 10.1109/IEDM.2018.8614664

[61] X. Sun, S. Yu, "Impact of non-ideal characteristics of resistive synaptic devices on implementing convolutional neural networks," *IEEE Journal of Emerging Selected Topics in Circuits Systems*, vol. 9, no. 3, pp. 570–579, 2019. doi: 10.1109/JETCAS.2019.2933148

[62] T. Gokmen, W. Haensch, "Algorithm for training neural networks on resistive device arrays," *Frontiers in Neuroscience*, vol. 14, p. 103, 2020. doi: 10.3389/fnins.2020.00103.

[63] S. Huang, X. Sun, X. Peng, H. Jiang, S. Yu, "Overcoming challenges for achieving high in-situ training accuracy with emerging memories," *IEEE/ACM Design, Automation & Test in Europe Conference (DATE)*, pp. 1–4, 2020. doi: 10.23919/DATE48585.2020.9116215

[64] S. Ambrogio, P. Narayanan, H. Tsai, R.M. Shelby, I. Boybat, C. Di Nolfo, S. Sidler, M. Giordano, M. Bodini, N.C.P. Farinha, B. Killeen, C. Cheng, Y. Jaoudi, G.W. Burr, "Equivalent-accuracy accelerated neural network training using analogue memory," *Nature*, vol. 558, pp. 60–67, 2018. doi: 10.1038/s41586-018-0180-5

[65] X. Sun, P. Wang, K. Ni, S. Datta, S. Yu, "Exploiting hybrid precision for training and inference: a 2T-1FeFET based analog synaptic weight cell," *IEEE International Electron Devices Meeting (IEDM)*, 2018, pp. 3.1.1–3.1.4. doi: 10.1109/IEDM.2018.8614611

[66] Q. Dong, M.E. Sinangil, B. Erbagci, D. Sun, W.-S. Khwa, H.-J. Liao, Y. Wang, J. Chang, "A 351 TOPS/W and 372.4 GOPS compute-in-memory SRAM macro in 7nm FinFET CMOS for machine-learning applications," *IEEE International Solid-State Circuits Conference (ISSCC)*, 2020, pp. 242–244. doi: 10.1109/ISSCC19947.2020.9062985

[67] C.-X. Xue, T.-Y. Huang, J.-S. Liu, T.-W. Chang, H.-Y. Kao, J.-H. Wang, T.-W. Liu, S.-Y. Wei, S.-P. Huang, W.-C. Wei, et al., "A 22nm 2Mb ReRAM compute-in-memory macro with 121-28TOPS/W for multibit MAC computing for tiny AI edge devices," *IEEE International Solid-State Circuits Conference (ISSCC)*, 2020, pp. 244–246. doi: 10.1109/ISSCC19947.2020.9063078

[68] N.P. Jouppi, C. Young, N. Patil, D. Patterson, G. Agrawal, R. Bajwa, S. Bates, et al., "In-datacenter performance analysis of a tensor processing unit," *ACM/IEEE International Symposium on Computer Architecture (ISCA)*, 2017, pp. 1–12. doi: 10.1145/3079856.3080246

[69] S. Yu, X. Sun, X. Peng, S. Huang, "Compute-in-memory with emerging nonvolatile-memories: challenges and prospects," *IEEE Custom Integrated Circuits Conference (CICC)*, 2020, pp. 1–4. doi: 10.1109/CICC48029.2020.9075887

[70] Y. Long, X. She, S. Mukhopadhyay, "Design of reliable DNN accelerator with unreliable ReRAM," *IEEE/ACM Design, Automation & Test in Europe Conference (DATE)*, 2019, pp. 1769–1774. doi: 10.23919/DATE.2019.8715178

Index

Page numbers in *italics* refers figures.

1.5T split gate, 93
193 nm photolithography, 46
1T1C, 55, 174
1T1R, 138, 153, 158, 166
2.5D and 3D heterogeneous integration, 74
3D X-point, 154
6T cell, 21
6T cell layout, 44
8T cell, 36

A

air gap, 69, 119
area efficiency, 7
Arrhenius plot, 113, 151, 158, 164, 174
aspect ratio (AR), 68
asymmetry/nonlinearity, 184

B

bank, 53
beta ratio (β), 24
bias temperature instability (BTI), 41
bit line (BL), 5
bit-cost-scalable (BiCS), 123
block, 96
body factor (m), 37
buried gate, 72
butterfly curve, 27

C

cache, 1
cell-to-cell interference, 117
Channel hot electron (CHE), 91
channel inhibition, 99
channel self-boost, 99
charge pump, 102
charge sharing, 56
charge-trap transistor, 120
chiplet, 74
CMOS under the array (CuA), 129
coercive field (E_c), 170
common plate (CP), 56
compute efficiency (TOPS/mm^2), 187
compute-in-memory (CIM), 180
conductive bridge RAM (CBRAM), 145, 155
confined cell, 149
contact gate pitch (CGP), 15
contact poly pitch (CPP), 15
critical charge amount (Q_{crit}), 32
cross-point array, 139

D

decoder, 7
DNN+NeuroSim, 183
double-data-rate (DDR), 53
drain coupling ratio α_{CG}, 88
drain-induced-barrier-lowering (DIBL), 48
DRAM layout, 66
DRAM refresh, 58
DRAM scaling trend, 66
DRAM sense margin, 57
DRAM timing, *60*
driver, 8
dual in-line memory module (DIMM), 53
dynamic noise margin, 32

E

EEPROM, 83
embedded DRAM (eDRAM), 78
embedded Flash (eFlash), 92
embedded memories, 4
energy efficiency (TOPS/W), 186
equalizer, 22, 58
equivalent bit area (F^2), 6
equivalent oxide thicknesses (EOT), 68
extreme ultraviolet (EUV) lithography, 46

F

fatigue process, 172
feature size, 13
ferroelectric field-effect transistor (FeFET), 176
ferroelectric random access memory (FeRAM), 174
Ferroelectrics, 169
Ferroelectrics endurance, 172
few electrons problem, 119
field switching MRAM, 161
fin depopulation, 15
FinFET, 18, 46
Flash disturb, 115
Flash endurance, 111
Flash retention, 113
Flash translation layer (FTL), 103
floating body cell, 79
floating-gate transistor, 85
F-N tunneling, 89
folded bit line, 65
foundry, 5

195

G

gain cell, 80
gamma ratio (γ), 26
gate coupling ratio α_{CG}, 87
gate length, 13
gate-induced-drain-leakage (GIDL), 38
gate-last process, 125
Ge-Sb-Te (GST), 147

H

$Hf_xZr_{1-x}O_2$ (HZO), 171
high-bandwidth memory (HBM), 76
high-k dielectric, 69
high-k/metal gate, 18
hybrid bonding, 76

I

imprint effect, 173
incremental step pulse programming (ISPP), 109
input/output (I/O) interface, 53
insulator-metal-transition (IMT), 145
intrinsic parameter fluctuations, 40
IR drop, 141

L

Landau theory, 170
latch-up effect, 43
level shifter, 102
line edge roughness (LER), 41

M

M1 pitch, 15
M1 track, 16
magnetic random access memory (MRAM), 159
magnetic tunnel junction (MTJ), 159
main memory, 2
mat, 53
memory hierarchy, 1
micro-bump, 75
Moore's law, 12
multi-level cell (MLC), 107, 149, 155, 177
mushroom cell, 149
mux, 8

N

N curve, 30
NAND array, 96
non-volatile memory (NVM), 2
NOR array, 93
NVMe, 85

O

one-time programmable (OTP) memory, 83
open bit line, 65
orthorhombic phase, 171
Ovonic-threshold-switch (OTS), 145
oxide RRAM (OxRAM), 155

P

page, 96
pass-gate (PG), 21
PCM drift, 150
PCM endurance, 152
PCM retention, 151
PCM thermal cross-talk, 151
perpendicular MTJ, 165
phase change memory (PCM), 147
positive-up-negative-down (PUND) testing, 172
Preisach model, 171
program/erase, 89
pull-down (PD), 21
pull-up (PU), 21

Q

quadruple-level cell (QLC), 107

R

random dopant fluctuation (RDF), 40
random telegraph noise (RTN), 41
read-modify-write (RMW), 36
recess channel, 71
remnant polarization (P_r), 170
resistive random access memory (RRAM), 155
row hammer effect, 63
RRAM endurance, 156
RRAM retention, 157

S

SD card, 84
selector, 139, 144
sense amplifier (SA), 9
separatrix, 33
Shmoo plot, 34
single-event-upset (SEU), 43
single-level cell (SLC), 107
sneak path, 141
solid-state-drive (SSD), 2, 84
source-side injection (SSI), 93
spin-orbit-torque (SOT) MRAM, 168
spin-transfer-torque (STT) MRAM, 162
SRAM hold, 21
SRAM read, 22

SRAM read timing, 24
SRAM scaling trend, 17, 45
SRAM write, 25
SRAM write timing, 26
stacked capacitor, 64
stacked nanosheet, 19
staircase WL contact, 125
standalone memories, 4
static noise margin (SNM), 27
storage node (SN) capacitor, 56
strained silicon, 17
stress-induced leakage current (SILC), 113
string select line (SSL), 97
STT-MRAM endurance, 164
STT-MRAM retention, 164
sub-array, 5
sub-threshold slope (SS), 37

T

technology node, 13
thermal stability factor (Δ), 164
through-silicon-via (TSV), 74
total-ionizing-dose (TID), 43
trench capacitor, 63
triple-level cell (TLC), 107
tunneling magneto-resistance ratio (TMR), 160

U

USB memory, 84

V

vector-matrix multiplication (VMM), 180
vertical channel 3D NAND, 122
volatile memory, 2
voltage transfer curve (VTC), 27

W

wake-up process, 172
word line (WL), 5
work function variation (WFV), 41
write back, 57, 175
write-and-verify, 109

X

X-stacking, 131